Mapping an Atlantic World, circa 1500

T0195305

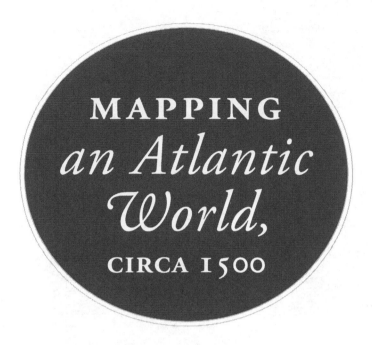

MAPPING
an Atlantic World,
CIRCA 1500

ALIDA C. METCALF

JOHNS HOPKINS UNIVERSITY PRESS
Baltimore

© 2020 Johns Hopkins University Press
All rights reserved. Published 2020
Printed in the United States of America on acid-free paper

2   4   6   8   9   7   5   3   1

Johns Hopkins University Press
2715 North Charles Street
Baltimore, Maryland 21218-4363
www.press.jhu.edu

Library of Congress Cataloging-in-Publication Data

Names: Metcalf, Alida C., 1954– author.
Title: Mapping an Atlantic world, circa 1500 / Alida C. Metcalf.
Description: Baltimore : Johns Hopkins University Press, 2020. |
Includes bibliographical references and index.
Identifiers: LCCN 2019056150 | ISBN 9781421438528 (hardcover) |
ISBN 9781421438535 (ebook)
Subjects: LCSH: Atlantic Ocean Region   History   16th century. |
Cartography—Atlantic Ocean Region—History—16th century. |
Cartographers—History—16th century.
Classification: LCC GA368 .M48 2020 | DDC 526.09182/109031—dc23
LC record available at https://lccn.loc.gov/2019056150

A catalog record for this book is available from the British Library.

*Special discounts are available for bulk purchases of this book. For more information,
please contact Special Sales at specialsales@press.jhu.edu.*

Johns Hopkins University Press uses environmentally friendly book materials,
including recycled text paper that is composed of at least 30 percent post-consumer
waste, whenever possible.

To Mary Clare, Van, and Michael

CONTENTS

COLOR PLATES

FIGURES

ACKNOWLEDGMENTS

This book grew out of my fascination with Brazil and how it was first mapped. The *Carta del Cantino* of 1502 first attracted my attention. Not only does this large, hand-painted and hand-lettered chart present a striking view of the world, but beautiful illustrations of parrots and trees illuminate Brazil. Five years later, Martin Waldseemüller created a huge printed world map known as *Universalis Cosmographia* that also included a remarkable view of Brazil. In the long and undulating South American continent, Waldseemüller placed a parrot but, of more significance, he entered the name "America." What inspired these choices? And how did this mapping of Brazil fit into the larger history of the opening of the Atlantic World? These questions made clear how important maps are for historians yet also how unequipped we are for understanding them. Familiar but often inscrutable, maps are packed with historical and cultural information. This book—the result of my scholarly journey as a social and cultural historian into the world of historical maps—is written for those interested in how maps first represented the Atlantic World in 1500.

I first encountered the *Carta del Cantino* and Waldseemüller's *Universalis Cosmographia* when writing *Go-betweens and the Colonization of Brazil*. In considering how these two historical maps represented Brazil, I realized that the mapmaker is a kind of representational go-between. In the typology presented in that book, the representational go-between chronicles, synthesizes, and represents to others. Most typically seen as writers, among the representational go-betweens, I concluded, are the makers of maps. This book began to take shape as I turned to other early maps of the Atlantic World and considered how the mapmaker, through their imagery and texts, explained to viewers what they were seeing. Their maps provided a visual analogy, supplemented with text, of a new world that could be encountered by sailing the Atlantic Ocean. The final form of this book grew out of a paper I gave at the Conference of Latin American History at the American Historical Association Annual Meeting (2016) on a session on the Atlantic World. I thank Jane Landers, who organized the session, as well as others at it, for their perceptive comments.

There are many whom I wish to thank for their advice, encouragement, and enthusiasm. First and foremost, I thank librarians and archivists who have made it a priority to digitize in high resolution their map collections. I was able not only to see nearly all of the maps discussed in this book but also to study them at leisure and in detail using digital reproductions. At Trinity University in San Antonio, I taught a first-year seminar on Amerigo Vespucci and the world chart sometimes attributed to him known variously as the *Portolan Weltkarte, Four Finger World Chart*, the *Kunstmann II*, or—as I shall refer to it in this book—*Weltkarte*. I thank my students for their enthusiasm and creativity, and I especially thank the Coates Librarians Michelle Millet, Jeremy Donald, and Diane Graves for their technical support. While still at Trinity, I received a Luso-American Foundation award for research in Lisbon; there I worked at the Arquivo Nacional da Torre do Tombo researching the Portuguese chartmaker Pedro Reinel. A Helm Fellowship at the Lilly Library at Indiana University allowed me to study its excellent collection of sources on early Americana.

At Rice University, where the book was researched and written, I thank Farès el-Dahdah for always encouraging me to finish even as the *imagineRio* project loomed with exciting new possibilities. The Fondren Library supported this project in more ways than I can properly acknowledge. Lisa Spiro, executive director of Digital Scholarship Services, has been a constant source of new ideas for scholarship using digital tools. Kim Ricker and Jean Aroom at the GIS Data center taught me ArcGIS. Jane Zhao at the Digital Media Commons introduced me to Zotero. I am also grateful to Anna Shparberg, the humanities librarian, and Jet Prendeville, the art/architecture librarian, for building our collection in historical cartography. At the Woodson Research Center, Amanda Focke, Dara Flinn, and Norie Guthrie assisted me on many occasions. Monica Rivero taught me the basics of image databases and metadata. My colleagues in art history, Diane Wolfthal and Linda Neagley, sharpened my appreciation of the aesthetic dimensions of historical maps, while my colleague Tani Barlow introduced me to the concept of ephemera. In the History Department, I thank Jim Sidbury, Alex Byrd, Aysha Pollnitz, Randal Hall, Caleb McDaniel, Fay Yarbrough, and Daniel Domingues da Silva for their many perceptive insights into the Atlantic World. Graduate students at Rice, particularly Wright Kennedy, Joice Oliveira, Sean Smith, Ludmila Maia, Rachael Pasierowska, Wes Skidmore, Livia Tiede, and Miller Wright shared their passion for Atlantic history. In the History Department, I am grateful to Beverly Konzem, department administrator, who untangled complicated questions related to travel and research, and to Erin Baezner, department coordinator, for assistance with scanning and printing. A teaching release from Rice's Humanities Research Center in 2014, as well as my chair's leave in 2018, allowed me to focus on research and writing.

My colleagues at Unicamp, especially Silvia Lara, Lucilene Reginaldo, Ricardo Pirola, and Bob Slenes provided many critical perspectives on the Luso-Atlantic world. Colleagues from Portugal, in particular Francisco Contente Domingues and Joaquim Alvares Gaspar, were essential as the project began, and I also wish to thank Monica Bello and Felipe Castro for their inspiration and encouragement. I am grateful to John Hessler, who invited me to view the Waldseemüller *Universalis Cosmographia* before it went into its special case at the Jefferson Building at the Library of Congress. I also thank Chet Van Duzer for many stimulating presentations, especially his excellent lecture at Rice on the Martellus *mappamundi*. The innovative Commission on Cartographic Heritage into the Digital (formerly known as Digital Approaches to Cartographic Heritage of the International Cartographic Association), and in particular their international conferences at The Hague (2011), Rome (2013), and Venice (2017), showed me exciting new ways to use digital maps in historical research. I thank Evangelos Liveratos for encouraging me to explore what geographic information systems can offer historians. I benefited greatly from attending two conferences organized by the International Conference on the History of Cartography presented in conjunction with the journal *Imago Mundi*, the first at Antwerp (2015) and the second in Amsterdam (2019).

For the final stages of the manuscript preparation, I wish to thank Jean Aroom and Amy Ferguson, who helped me to conceptualize several of the figures that appear in this book. Rice Architecture students Andrea Rubero and Christina Zhou created the final, publication-quality figures. Mario Norton in Fondren Library's Digital Media Commons generated the images of Behaim's and Waldseemüller's globes. I thank my dean, Kathleen Canning, and my chair, Carl Caldwell, for supporting the permissions needed to reproduce a dozen maps in color. I thank the editorial team at Johns Hopkins University Press for its careful work on the final manuscript.

I am deeply appreciative of my friends and family. Joan Burton gave the first chapters a critical read and clarified Latin inscriptions on maps. Janis Arnold has been an inspirational source on the craft of writing. My husband, Daniel Rigney, read an earlier draft manuscript with great care. My sons, Matthew and Benjamin, increasingly followed from afar the progress of this book, but they never lost their curiosity about the story I was seeking to tell. My late parents and my husband's late parents did not live to see the publication of this book, and because they were all so fascinated by it, I regret that I was unable to finish more quickly. I dedicate the book with great affection to my siblings, my oldest and dearest traveling companions.

Almost all of the maps described in this book are available as high-resolution digital copies that can be consulted online through the websites of the libraries, archives, and museums that hold them. I encourage readers to access these maps in order to appreciate their stunning detail and color. Information on each map, where it is located, and its web address (as of this writing) are given in the notes and bibliography. Maps are often known by several names, and in the following pages I refer to maps by the names given to them by the libraries and archives that hold them. Legends from maps, as well as quotations from other primary sources, appear in the text in English, but the original language is available wherever possible in the notes.

Mapping an Atlantic World, circa 1500

Introduction

There was a moment in historical time when the Atlantic Ocean moved from the periphery to the center on European world maps. That event dates to 1500, and it marks not only a paradigm shift in how the world was mapped but also the opening of what historians call the Atlantic World. By examining the earliest surviving maps, I seek to understand how the first depictions of the entire Atlantic Ocean carried powerful and persuasive arguments about the possibility of an interconnected Atlantic World.

There is a profound difference between the history of the Atlantic Ocean and the history of the Atlantic World. The Atlantic Ocean is a geographic entity that has its own history that must be told in the vast epochs of geologic time. Historians developed the concept of the Atlantic World in order to explain the consequences of extensive and sustained movement of "people, plants, pathogens, products, and cultural practices" across the Atlantic Ocean after 1500. The moment when the Atlantic World first appeared on historical maps—1500—is hardly significant in the history of the Atlantic Ocean. However inconsequential for the Atlantic Ocean, this moment is important to historians, for it explains dramatic change and sustained new continuities. After 1500, when peoples, animals, plants, commodities, and pathogens crossed the Atlantic Ocean through commerce, colonial enterprises, and religious missions, these transatlantic circulations—whether economic, demographic, ecological, or cultural—affected peoples living along the Atlantic coasts and rural hinterlands of Africa, the Americas, and Europe. Although local conditions were distinctive and unique, transoceanic movements created common patterns and experiences that linked those living in, around, or from the Atlantic Ocean.[1]

Oddly, often missing from histories of the Atlantic World is the Atlantic Ocean itself. While recognizing the compelling histories written about the Atlantic World, historian Alison Games notes that "the Atlantic history that many historians produce is rarely centered around the ocean, and the ocean is rarely relevant to the project." The moment when the Atlantic Ocean came to be seen as a geographic entity to be explored and exploited was something new and highly significant, according to historian N. A. M. Rodger. By the mid- to late sixteenth century, the Atlantic Ocean had emerged as a geographic entity, and by the eighteenth century the label "Atlantic Ocean" was pervasive on maps.[2]

Perhaps one reason for the omission of the ocean from Atlantic World history is that it seems so obvious that it hardly requires explanation. The Atlantic Ocean's emergence on charts, maps, and globes appears to be such a logical step in the evolution of more accurate cartography that it requires little reflection. As new areas are discovered, this line of reasoning holds, implicitly they are mapped. Therefore, historical maps will necessarily reveal the expanding knowledge of the Atlantic Ocean, and maps will change accordingly. In other words, as the ocean was explored, its mapping was inevitable.[3]

There is much truth to this premise—that maps follow and record knowledge—but maps are complex historical documents. The eminent scholars of historical cartography J. B. Harley and David Woodward define maps as "graphic representations that facilitate a spatial understanding of things, concepts, conditions, processes, or events in the human world." In this vein, maps "represent"—they do not simply mirror—and they "facilitate"—meaning that they interpret. In other words, maps structure as they document; they shape as they present. Cultural historian Christian Jacob describes the map as a "mediation" between the cartographer and the map viewer. He sees the map as an instrument that transfers and translates, allowing "the transmission of knowledge and the confrontation of different experiences." These definitions emphasize that maps are rich textual and visual sources; maps do not simply record geography.[4]

Similarly, the crossing of the Atlantic Ocean—and its importance in the history of Western Europe, Africa, and the Americas—is so familiar that it too seems to require little further explanation. Notwithstanding the excavations at L'Anse aux Meadows in the 1960s, which document early Norse settlements in eastern Canada, the usual story of the opening of the Atlantic Ocean still begins with 1492. Once Columbus returned, the resulting Spanish conquests—first of Hispaniola, which created the model for Cuba, which then provided the staging ground for Mexico—are understood as predictable. Traditional explanations for what motivated small bands of men—their desires for gold, glory, and God—have long been persuasive. Even when modified by the powerful arguments introduced by Alfred Crosby in the 1960s—that biological and ecological factors largely explain the Spanish conquest of Mexico and Peru—the result is still seen as unavoidable. Jared Diamond's insistence that guns, germs, and steel tilted the encounter to favor the Europeans has reinforced the widespread acceptance that the fall of indigenous societies in the Americas was inescapable.[5]

Such rapid and profound change at the turn of the sixteenth century, when in a few short years the crossing of the Atlantic came to be seen as routine, requires more explanation. Why was so much invested so quickly in exploring, developing, and risking lives and fortunes in the distant lands of the western Atlantic? So many of the supposedly inexorable events depended on new and startling decisions, such as the willingness to cross the ocean, the visualization of sources of wealth in places known only recently, the desire to create religious missions in unfamiliar regions, the belief that colonies could thrive at great distance, or the eagerness to invest in unpredictable and untested enterprises. This book argues that an important piece of the explanation lies with early maps that offered a vivid new image of the Atlantic Ocean and persuasively promoted an interconnected Atlantic World.

As of this writing in 2020, we appear to be on the brink of another rapid reappraisal of the Atlantic Ocean and its role in the world. Although we know the Atlantic as the second-largest ocean in the world, covering 20 percent of the earth's surface and comprising an area of approximately 41,105,000 square miles, much of what we know about the Atlantic will be affected by global climate change. As the ocean temperatures rise, weather patterns as well as currents, winds, and food chains will be impacted. As sea levels creep upward, the historical ports of the Atlantic World, many of which have developed into the major cities of the modern world—New York, London, Amsterdam, Rio de Janeiro, Buenos Aires, Lagos, Cape Town, and many more—will have to adapt quickly to remain viable cities. Just as a new idea of the nature of the Atlantic changed rapidly in the years before and after 1500 and ushered in profound change, so too will our understanding of the Atlantic be transformed, as will be the lives of the modern descendants of the historical Atlantic World.[6]

This book, then, is not a history of the Atlantic Ocean; instead it is a close examination of a moment in historical time when the Atlantic World opened. Cartographic historian Matthew Edney has noted that the "Atlantic has never been a natural, predefined stage on which humans have acted: like all spatial entities, it is a social construct that has been constituted through human activities." Following his insight, I have looked at maps to see how mapmakers charted the Atlantic Ocean and proposed an Atlantic World. The depiction of the Atlantic Ocean on historical maps was largely that of the surface—the coastlines, islands, reefs, and shoals. The mapping of the ocean itself—its depths, currents, ecological zones, and plant and animal life—would come much later in time, and in fact it is ongoing to this day.[7]

Two rich fields of scholarly inquiry underly the approach taken in this book. The first is that of the Atlantic World. The concept of the "Atlantic World," as defined and unpacked by Bernard Bailyn, is the contention that Western Europe, West Africa, and the Americas can be studied as an integrated unit from 1500 to 1800. As Bailyn illustrates, the concept has its own history, from the idea of an "Atlantic Community"—first coined in 1917—to "Atlantic History," which became increasingly influential after the Second World War. For Bailyn, the approach emerged gradually from the work of historians as they followed "the logic of the subject as it unfolded." The geographer D. W. Meinig made the key intervention in his view by rejecting the analogy of European discovery in favor of "a sudden and harsh encounter between two old worlds that transformed both and integrated them into a single New World."[8]

Historians debate the idea and the importance of the Atlantic World, but many have found its transnational, comparative, and interdisciplinary focus revealing. For economic historians, the emergence of cash-crop agricultural economies based on tropical commodities such as sugar, indigo, tobacco, coffee, cotton, or cacao is a key phase in the development of a global economy. For social historians, the movement of peoples back and forth across the Atlantic and the impact of such movement on family forms present a compelling chapter in historical demography. Historians assign great importance to the forced migration of upward of ten million Africans, enslaved in their homelands, carried across the Atlantic under terrifying shipboard conditions, and sold in port cities across the Americas. This event not only was a violent and tragic story in modern history but was central to the social, economic and cultural development of the Americas. A smaller slave trade moved indigenous peoples across the Atlantic in the sixteenth century, while an internal slave trade forcibly relocated native men, women, and children within the Americas. Historians of religion have always been attentive to how local religious traditions incorporated traditional pre-Christian beliefs, but an Atlantic World framework allows them to study common patterns in the spread of Christianity in Africa and the Americas. Environmental historians have studied not only how agricultural practices and disease environments on both sides of the Atlantic changed as unknown diseases caused a demographic collapse in the Americas in the sixteenth century but also how hoofed animals brought a radical transformation of the landscape. Cultural historians have emphasized the importance of hybridity, which appeared throughout the Atlantic World, as did the construction of new identities, whereas political historians have long been attentive to the spread of ideas throughout the Atlantic during the late eighteenth- and early nineteenth-century Age of Revolution.[9]

Atlantic World history extends broadly, in both chronology and spatial extent, and it moves beyond the study of the colonial empires established by European powers. Historian Jorge Cañizares-Esguerra notes that "the many peoples that lived on the Atlantic basin were connected to countless other communities outside the formal boundaries of empire." For example, in the case of Atlantic commerce, whatever the commodity traded, it generated "commercial and ethnic entanglements that rendered the entire Atlantic basin into a large borderland of porous boundaries." Central to the study of the Atlantic World is the slave trade and the recognition that it must be understood transnationally. The experiences of individual slaves, slave families, and slave communities cannot be comprehended solely through the lens of colonial or national histories. Similarly, the Columbian Exchange paid no heed to the political boundaries of empires. Foods native to the Americas, such as corn, beans, potatoes, and manioc, became essential to the subsistence strategies of small farmers in Africa and Europe, while plants native to Europe and Africa spread widely throughout the Americas. In the nineteenth century, political changes, such as the American Revolution, the Haitian Revolution, and the Spanish-American Independence movements, broke apart the former colonial empires, while the abolition of the slave trade ended the brutal forced migration of hundreds of thousands of Africans. For historians of the Atlantic World, even as these events have great significance for the creation of nation-states, their transnational—Atlantic—roots must be acknowledged.[10]

What this book contributes to the field of Atlantic history is the importance of the moment—1500—and the essential contribution made by mapmakers. Historical maps have always been important in histories written about the Atlantic World, yet rarely have historians placed them at the center of their analyses. Historical maps often appear as passive background illustrations, whereas the maps created for publication are more active because they convey the arguments made by the author. In this book, historical maps appear in the foreground and are analyzed as both geographic and cultural sources. The goal is to see how maps conveyed knowledge about the Atlantic Ocean and how they projected the possibility of an integrated Atlantic World.[11]

As suggested recently by Peter Barber, the study of historical cartography has three major developmental phases: the traditional, the internal, and the sociocultural contextual. Traditional approaches emphasized the search for surviving maps and the identification of the mapping traditions from which they came, as well as the roles of major cartographers. The resulting publications were carto-bibliographies, often illustrated, that grew out of studies of map collections. Such carto-bibliographies presented the known maps from national "schools" of mapping—the Portuguese, Spanish, Dutch, French, British, German, and so on—and they illustrated the evolution of mapping traditions over time. Volumes organized around a theme or region—such as the Americas—presented reproductions of historical maps with short descriptions of their importance.[12]

Barber associates the internal approach used by historians of historical cartography with historian David Woodward, and this methodology emphasizes a fuller study of each map: who made it, when and where, its medium, its symbology, its sources, and its moment in time. This approach insists that the historian must first establish the basic parameters of the map before attempting to interpret it.

This approach is fully evident in the monumental publication by the University of Chicago Press: *The History of Cartography*. The project builds on the foundation laid by traditional scholars, as can be seen in the organization of sections where key essays treat national mapping traditions while also providing synthetic, comprehensive, and detailed analysis of individual maps.[13]

The third approach identified by Barber developed out of the work of J. B. Harley, who transformed the field of historical cartography with a series of papers and articles that argued for the importance of understanding the political and cultural messages imbedded in each and every map. Harley challenged historians of historical cartography to uncover the ways in which maps served powerful interests. He argued for attention to how maps shaped perspectives of places by noting the "silences" imbedded in maps, a technique through which they can privilege or exclude information. Furthering these arguments, Denis Wood argues in *The Power of Maps* that maps are complex textual images with internal and external codes. These codes structure both how the map is made and how the map is interpreted.[14]

Even as scholars of historical cartography have studied maps from a variety of perspectives and in great detail, they well know that maps—as primary sources—are inherently problematic. This is particularly true for this study of the opening of the Atlantic World. There were not many maps, globes, and charts made in in the years immediately before and after 1500, and even fewer survive to our times. Maps are fragile, transitory documents, and many, if not most, eventually crumble or are thrown away. This is true whether maps are hand drawn, printed on paper, or created digitally; all are potentially fleeting. An essential quality of maps—the need to constantly update them—underlies their ephemerality. Once out of date or worn out, maps are discarded. Of historical maps, only the more elaborate ones or those either bound into books or collected by libraries have survived. Many of the individual maps that have lasted to the present, according to David Quinn, were also valued as works of art.[15]

The earliest surviving maps and charts to show the Atlantic Ocean are the primary sources examined in this book. A key distinction will be made between a map and a chart. Unlike maps, which depict land and emphasize features of the landscape, charts focus on the sea. Used broadly, the term "map" can encompass a chart, as well as other plans and diagrams, but a chart has a specific use—navigation—and it privileges the information needed by sailors. Modern charts are used to set courses and to determine where a craft is; they contain measurements of water depths, underwater hazards, currents, lighthouses and buoys, a scale, latitude and longitude, and a compass arrow pointing north. Even as sailors navigate with the Global Positioning System, or GPS, with satellite connections that pinpoint the exact location of their vessel, many still carry paper charts. The origins of modern charts—whether paper or digital—lie in the historical charts of the Mediterranean and Atlantic.[16]

This book takes advantage of all three approaches used by scholars of historical cartography and adapts them in order to study the opening of the Atlantic World. Many of the charts and maps that have been traditionally studied as part of national schools of cartography will be approached here in a national and transnational context. Chartmakers worked in ports such as Lisbon and often in the employ of the king, but they nevertheless had extensive contact with princes, sailors, merchants, and sea captains from many homelands. Information about the Atlantic circulated widely among mariners and merchants, and the sharing and copying of charts occurred, despite official attempts to limit the spread of cartographic knowledge. Similarly, the makers of world maps—who did not necessarily live in port cities—had many patrons, and they too took advantage of many sources of geographic and historical information that extended far beyond the kingdom, duchy, or republic where they lived. Printed in Latin, their maps could be read by a wide audience of educated humanists throughout Western Europe.

The in-depth internal study favored by Woodward is key to this book because its comprehensive and detailed approach is essential for understanding the early maps and charts of the Atlantic. Maps and charts can be compared in order to see differences between cartographic traditions at a specific moment in time. However, the evaluation of the geographic accuracy of the maps and charts will not be a focus; instead, what is more important is how the Atlantic was being conceptualized, not how accurately it was being mapped. Nor will attention be paid to the mathematical functions that underlay the making of historical maps and charts, for this approach, while indispensable to understanding historical navigation and the evolving accuracy of mapmaking, is beyond the scope of this book. The sociocultural approach advocated by Harley, who was a careful map historian, can result in the reading of maps solely as texts while losing sight of their fundamental geographic character and navigational uses. Although the approach taken in this book is closest to Harley's, every effort has been made not to lose sight of the essential geographic nature of maps and the navigational potential of charts.

The following chapters unfold first chronologically and then thematically. Chapter one illustrates that the Atlantic Ocean hardly appeared on classical and medieval maps. Rather, it was subsumed within the great Mare Oceanus (Ocean Sea) that circled the continents of Africa, Asia, and Europe. Even as the size of the Atlantic Ocean increased on late medieval maps and charts, it still remained peripheral. Chapter two reveals that a new view of the Atlantic Ocean appeared quite suddenly, but not immediately following 1492. Rather, the year 1500 was the tipping point. As those creating charts and maps reconceptualized the Atlantic and its place in world, they created striking and vast oceanic spaces that presented the first inklings of an Atlantic World.

Chapters three and four explore the techniques used by chartmakers, cosmographers, and artists. Chartmakers are the focus of chapter three. Not only did chartmakers first map the Atlantic Ocean—paying attention to coastlines and their hidden dangers, such as shoals and reefs—but they also included navigational information on their charts. The compass rose, rhumb lines, and a scale all aided in navigation and shaped how the surface geography of the ocean would be understood. Illuminations painted onto the chart, such as of cityscapes, animals, indigenous peoples, local rulers, and flags, also configured the understanding of the lands bordering the ocean. Whether creating a plain chart to be used on board ship for navigational purposes or a heavily illustrated chart intended for a powerful lord to contemplate, chartmakers visualized more than the Atlantic Ocean, they began to present an Atlantic World.

Atlantic charts influenced the work of cosmographers—the name given to Renaissance geographers—who turned to print, as we shall see in chapter four. Educated as humanists, cosmographers lived scattered throughout Western Europe and were interested in mapping the world and the heavens. During the Renaissance, artists were commissioned to create perspective city views, often drawn using grids, and to paint elaborate maps that transformed the walls and hallways of palaces. As cosmographers and artists incorporated the work of chartmakers into their maps, and as they used the medium of print, the new view of the Atlantic spread well beyond maritime communities. Martin Waldseemüller created a world map titled *Universalis Cosmographia* and a globe, both printed in 1507. It was Waldseemüller who first named the new mainland in the western Atlantic "America."[17]

The visual language that emerged for depicting the Atlantic is the theme explored in chapters five and six. Chartmakers first established iconographies for specific places in the Americas, and these visual codes derived from the eyewitness observations of sailors. As these visuals repeated from map to map, they tended to become more simplified and generic. Cannibalism, the violent and dominant trope used to depict native peoples in South America, is the subject of chapter six. Scenes of cannibalism cast the peoples living in specific places in the Atlantic World as uncivilized and cruel, and the repetition of such imagery on maps reinforced such negative representations and presented them as true.

Closing the book, the epilogue reflects on the ephemeral nature of these first charts and maps of the Atlantic Ocean and the Atlantic World. Fragile and fleeting, they nevertheless communicated in powerful and persuasive ways long after 1500.

CHAPTER ONE

The Atlantic on the Periphery

The first known maps that locate the Atlantic Ocean date to the time of the ancient Greeks, who called the inhabited world the *oikoumene* and believed it to be surrounded by water. In his histories, the Greek historian Herodotus referred to the sea outside the Straits of Gibraltar, past the familiar and known Mediterranean, as the Atlantic Ocean. At the narrow straits, the Greek mythological hero Heracles (later known to the Romans as Hercules) was said to have erected two pillars that marked the end of the known world. Just beyond the straits, and often associated with the pillars, lay the island of Gades off the coast of Spain. And beyond Gades lay the vast and largely unknown ocean.[1]

The waters of the boundless Atlantic Ocean inspired imagination. In his dialogue *Timaeus*, Plato has Critias tell a story, purportedly from his great-grandfather, about an epic battle between a mighty power in the Atlantic and Athens. Critias explains that nine thousand years earlier the Atlantic Ocean had been navigable and that a great kingdom—Atlantis—arose on an island beyond the Pillars of Heracles. After conquering North Africa and parts of Europe, Atlantis threatened the Athenians, but in an epic battle Athens defeated Atlantis. Afterward, Critias reports, an earthquake and flood caused Atlantis to sink, leaving behind muddy shoals that made the Atlantic impenetrable.[2]

How Greek maps portrayed the Atlantic Ocean must be reconstructed from texts, because the maps, charts, and globes have all been lost. According to Greek and Roman authors, Anaximander (ca. 611–546 BC) is believed to have been the first to make a world map and a globe. Anaximander's long-lost map would have been circular, and the Atlantic would have formed part of the ring of oceans that enclosed the *oikoumene*. Later maps and globes are known only from writings about them, such as those of Herodotus and Aristotle, who both had occasion to comment on maps they had seen. Brief asides about maps make clear that Herodotus and Aristotle found fault with how maps of their day depicted the oceans, which would have included the Atlantic. Both thought that maps wrongly represented the world by depicting it as circle with the oceans flowing around in an outer circumference.[3]

Romans learned geography and the art of mapmaking from the Greeks, and Roman writers were familiar with both the Atlantic Ocean and how it was mapped. Pliny the Elder devotes one chapter of his encyclopedic *Natural History*, begun in 77 CE, to lands that had been turned into oceans, explaining that "cases of land entirely stolen away by the sea are, first of all (if we accept Plato's story), the vast area covered by the Atlantic." There are other references to the Atlantic in Pliny's *Natural History*, and one, in the introduction to book III, spells out the place of the Atlantic on a map. "The whole circuit of the earth is divided into three parts, Europe, Asia and Africa," he begins. "The starting point is in the west, at the Straits of Gibraltar, where the Atlantic Ocean bursts in and spreads out into the inland seas." Pliny's text then suggests that he had a map before him as he wrote: "On the right as you enter from the ocean is Africa and on the left Europe, with Asia between them; the boundaries are the river Don and the river Nile."[4]

As the Roman Empire expanded, Greek geographers developed new techniques for representing the known world. Marinus of Tyre (ca. 70–130 CE) is credited with the design of a projection today known as the equirectangular projection (also known as the plate carrée or plane chart). A projection, as defined by John Snyder, is the "systematic representation of all or part of the surface of a round body, especially the earth, onto a flat or plane surface." In the equirectangular projection, all parallels of latitude are straight, equally spaced, parallel lines, and all meridians of longitude are likewise straight, equally spaced, and parallel. The result is a rectangular grid in which the lines of latitude and longitude intersect at right angles. On such a map, the world appears as flat, not round, and the Okeanos (Ocean) would have appeared in the West and in the East, with the continents of Europe, Africa, and Asia in between. As Marinus believed the Canary Islands marked the edge of the inhabited world in the West, they would have marked the extent of the Atlantic on a world maps created at the time using this projection. Much of what is known about Marinus comes from Klaudios Ptolemaios (ca. 90–168 CE), another Greek geographer living in the Roman Empire, known today by his Latin name, Claudius Ptolemy. The work of Ptolemy was rediscovered in the Renaissance, and in his *Geōgraphikē hyphēgēsis* (ca. 150 CE), known in English as *Geography*, Ptolemy summarized and commented on Greek knowledge of the world and provided instructions for making world maps. Ptolemy described three additional projections that mapmakers could use when creating maps. All of these placed the Atlantic in the West at the edge of the known world.[5]

Although mapmakers in medieval Europe continued to use classical texts (such as Pliny's) for information about specific places, Ptolemy's work was unknown to them and their maps returned to the circular form in which the Atlantic was but a small piece of the ocean flowing around the known world, depicted as a circle. Medieval maps of the world, known as *mappaemundi* (singular *mappa-*

mundi), served many purposes, but in the main, David Woodward writes, they "blended concepts of both time and space as a context for understanding the Christian life." The surviving *mappaemundi* number approximately 1,100, most of which are found in books as illustrations. Map historians point to at least four general types of *mappaemundi*. One common type, particularly relevant for understanding how the Atlantic Ocean was mapped, consisted of a tripartite schema, later named by historians as the "T-O." Medieval *mappaemundi* created using the T-O were circular and gave an outlying, marginal role to the oceans. The O represents the known world, with the unnamed Atlantic as part of the Mare Oceanum (Ocean Sea) that surrounds it. The T is the axis created by the Mediterranean Sea, as well as rivers such as the Nile or the Don. Within the map, Africa, Asia, and Europe are the three continents that make up the world, with Asia typically presented as the largest.[6]

The mapping of the oceans as a narrow, peripheral edge on T-O *mappaemundi* can be seen on the small *mappamundi* illumination in Isidore of Seville's *Etymologiae* (*Etymologies*), a medieval encyclopedia published in the seventh century. Republished nine hundred years after Isidore's death by a German printer in 1472, the T-O *mappamundi* illumination (discussed later in the chapter) introduces book 14 on Asia.[7]

In the *Etymologies,* Isidore explains in his entry *De oceano* that the ocean has its name "because it goes around the globe [*orbis*] in the manner of a circle [*circulus*]." Although the T-O form suggests that the earth is flat, Isidore makes it very clear that the earth is in fact a sphere surrounded by water. In explaining the relationship between the earth (*terra*), which he conceptualizes as a globe, and the ocean, he writes, "indeed, the Ocean that flows around it encompasses its borders in a circle."[8]

Isidore clarifies that sections of the ocean are known by specific, regional names. He writes, "also, the Ocean takes different names from nearby areas, such as . . . Atlantic." Following Pliny and other classical authors, Isidore connects Hercules to the Straits of Gibraltar. They are named, he states, "for Gades [i.e., Cadiz], where the entrance of the Great Sea [i.e., the Mediterranean] first opens from the Ocean. Hence, when Hercules came to Gades he placed pillars there, believing that the end of the lands of the world was at that place."[9]

Large medieval *mappaemundi* were covered with detailed information, but mapmakers gave little attention to the Atlantic. Few large *mappaemundi* survive, yet sources suggest that such maps did exist and that they were important. Among them were *mappaemundi* that hung in cathedrals. Such *mappaemundi* were not, as Woodward emphasizes, "snapshots of the world's geography at a given moment in time, but a blending of history and geography, a projection of historical events on a geographic framework." They narrated key events in the history of Christianity, and they relied on imagery to tell that story. Sources for locations mapped came from the Bible, but classical knowledge informed not only the underlying placement of continents but also many place names and legends. Mapmakers added contemporaneous information, often drawn from the itineraries of travelers. On *mappaemundi*, time was metaphorical, and space was exaggerated or minimized to emphasize symbolic and religious meanings. The center of the world lay at Jerusalem, and the Mediterranean is mapped as fully known and fundamental to the world. The Atlantic Ocean is, in contrast, part of the thin channel of water that rings the world. Also at the edges of the known world are the terrestrial paradise and the places where monsters live.[10]

The famous world map in the Hereford Cathedral, dated circa 1300, is a classic example of a large late medieval *mappamundi*. It measures 1.59 × 1.34 m (about 5 × 4 feet) and was painted on a single calfskin. Originally the *mappamundi* was mounted inside a wooden triptych with folding doors. When the doors were open, the map had the central space, while the insides of the two door frames held religious imagery that evoked the annunciation: the archangel Gabriel appeared on one, and the Virgin Mary on the other. The author of the map signed it: "Richard of Holdingham, or of Sleaford, who made it and set it out." The map is both religious and secular, and it is a compendium of knowledge drawn from classical as well as Christian traditions. The image of the world is circular in form, representing God's creation as a sphere, and above, in an added triangular space, Christ sits in judgment.[11]

Encircling the world on the Hereford *Mappa Mundi* are two circumferences; one contains the names of the cardinal directions (N, S, E, W), while the other records the winds. Inside both is a ring of water (the oceans) that circles the world, at the center of which is Jerusalem—shown as a walled, circular city. The enormous land masses of Asia, Africa, and Europa are compressed into the T-O form. Long rivers cut through the continents, and two enormous inland bodies of water—the Mediterranean and Black Seas—separate them. All parts of the map are filled with text and illustrations. Cities and towns are drawn as towers and castles. Events from the Bible—such as the landing of Noah's ark, the Tower of Babel, or the parting of the Red Sea—embellish their respective regions. The illustrations are not limited to Christian stories, as classical myths and legends are depicted, such as the labyrinth on Crete, the camp of Alexander the Great, and the Pillars of Hercules. Illustrations of animals, plants, and fish—some recognizable, others magical or monstrous—are inserted everywhere, but a special region is devoted to monstrous peoples, who are placed to the east of the Nile River at the edge of Africa. The terrestrial paradise lies at the top of the map. Depicted as an island surrounded by a wall and ringed with fire, it appears as inaccessible to ordinary humans.[12]

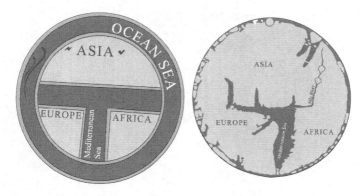

Fig. 1.1. The Atlantic on T-O *Mappaemundi*. Redrawn from *mappamundi*, ca. 603, in Isidore de Seville, *Etymologiae*, 1472 (*left*), and the Hereford *Mappa Mundi*, ca. 1300 (*right*).

In figure 1.1, the T-O *mappamundi* that appeared as a small illustration introducing book 14 of Isidore of Seville's *Etymologiae* is redrawn (figure 1.1, *left*) and placed next to a redrawing of the coastlines from the Hereford *Mappa Mundi* (figure 1.1, *right*). On Isidore's diagrammatic *mappamundi*, the Mare Oceanum (Ocean Sea) encloses the continents, while the Mediterranean separates Europe from Africa. Next to Isidore's *mappamundi* is a greatly simplified redrawing of the Hereford *Mappa Mundi*. The underlying T-O structure is clearly visible, and the same basic tripartite division of the known world (into Africa, Asia, and Europe) appears. Surrounding the known world is the thin band of ocean.

On the Hereford *Mappa Mundi*, the waters of the both the Mediterranean and the circular ocean appear dark and unmoving. The Pillars of Hercules mark the intersection between the Mediterranean and the Atlantic, and as the map is oriented with East at the top, West falls at the bottom, and it is here where Richard, the mapmaker, places two pillars on an island that he labels Gades. Next to them the legend reads "The Rock of Gibraltar and Monte Acho are believed to be the Columns of Hercules." Richard, following Pliny the Elder and Isidore of Seville, placed the pillars on the island named Gades, which was also the name that Pliny gave to the Straits of Gibraltar. On the map, the location of the Pillars of Hercules thus serves as a metaphor to mark the western boundary of the known world. Past the island on which Richard located the Pillars of Hercules, the unnamed Atlantic flows south along the coast of Africa and north along the coasts of Europe, as can be seen in the redrawing of this portion of the Hereford *Mappa Mundi* (figure 1.2).[13]

On the original Hereford *Mappa Mundi*, there is one slight hint of the name Atlantic, but the name does not refer to the ocean. In North Africa, near the legend for the Atlas Mountain, the mapmaker inserted the place name "Tlantica

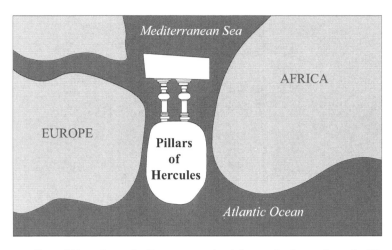

Fig. 1.2. Pillars of Hercules at the Entrance to the Atlantic. Redrawn from the Hereford *Mappa Mundi*, ca. 1300.

deserta." According to historian Scott Westrem, this "must be a corruption of 'Atlantica,'" and therefore Westrem translates the label as "The Atlantic Desert." In researching the sources used throughout the Hereford *Mappa Mundi*, Westrem finds references that suggest that both the region and the people who lived there had the name "Atlanticas" or "Atlantes."[14]

Mappaemundi drawn with the T-O form presented the Atlantic as a vague oceanic space bordering the known world, yet there was another mapping tradition in late medieval Europe that gave it far more prominence. This was the tradition of Mediterranean chartmaking. Whereas "the medieval *mappaemundi* are the cosmographies of thinking landsmen," map historian Tony Campbell writes, "by contrast, the portolan charts preserve the Mediterranean sailors' first-hand experience of their own sea, as well as their expanding knowledge of the Atlantic Ocean." Chartmakers emphasized coastlines, recorded names along coastlines, and illustrated major port cities. Although there is no consensus among scholars about the origins of these charts—often called portolan charts—the surviving examples make clear that they ranged from the simple and utilitarian to the richly illustrated. Among the illustrated charts were lavish and expensive world charts that gave the Atlantic more importance and detail than the *mappaemundi* made with the T-O form. Chartmakers generally did not use the T-O frame when making a *mappamundi* and instead used an underlying network of points and lines that marked the directions of the winds.[15]

The *Atlas de cartes marines*, also commonly known as the *Atlas Catalan*, is considered a *mappamundi* made in the chartmaking tradition. This world chart gives far greater significance to the Atlantic Ocean than a T-O world map, such as the Hereford *Mappa Mundi*. The presumed maker of the highly illuminated chart is Abraham Cresques, known to be a master compass and chartmaker in the Kingdom of Aragón. Dated 1375, the *Atlas de cartes marines* presents the world in a rectangular shape in twelve panels, and it combines knowledge commonly found on charts with information derived from classical and religious sources, as well as from the eyewitness accounts of travelers. The chartmaker included the travels of Marco Polo in Asia, as well as the wealth of the king of Mali, Mansa Musa, in Africa.[16]

In the chartmaking tradition, seas were more important than land, and this fact, in addition to the rectangular shape of the *Atlas de cartes marines*, causes the Atlantic to appear far larger than on a circular T-O *mappamundi*. The Atlantic is not a thin channel of water at the edge of the West. Instead it takes up much of one panel and even continues onto a second; it is mapped from West Africa in the south to the north of Scotland. As can be seen in figure 1.3, which is a redrawing of the coastlines on all panels of the world chart, the chartmaker has given the Atlantic Ocean a dramatically greater place in the world. Nevertheless, on the original world chart, the chartmaker still gave it a generic label: "Mare Ocheaenum" (Ocean Sea).

The chartmaker of the *Atlas de cartes marines* uses text and visual illustrations to present the Atlantic as a place beyond the Mediterranean where commercial exchanges, colonization efforts, and even missionary enterprises were taking place. The rectangular shape of the map provided more space for the Atlantic, but it was also the chartmaker's choice to present the Atlantic in greater detail. Beginning with the waters of the Atlantic, which are rendered in the same way as those of the Mediterranean—with thin, wavy, blue lines—the chartmaker suggests the movement between the two seas. Near the label "Mare Ocheaenum" appears a compass rose. The compass rose and the similar style used to paint the waters imply that skilled sailors could navigate from one body of water to the other and back again. A large illumination of a boat under sail off the coast of West Africa reinforces this

Fig. 1.3. The Atlantic on a Medieval World Chart. Redrawn from [Cresques], *Atlas de cartes marines*, 1375.

message. Flying a red and gold flag, which is a symbol of Cataluña, then part of the Kingdom of Aragón, the boat is headed for the West African shore to a place labeled "Cap de Finisterra" (Cape of the End of the Land). The textual legend next to the ship gives the ship's captain or owner's name as Jaime Ferrer, as well as the date, 1346, and the destination: riu de lor (River of Gold).[17]

Below the decorative compass rose, a long textual legend appears in the Atlantic at the edge of the *Atlas de cartes marines*. The legend describes the Canary Islands, and the commentary comes directly from Pliny the Elder and Isidore of Seville. The islands are described as fertile with tall trees and abounding in fruits and birds. As noted by historian Francesc Relaño, the Canary Islands, though known, had not been visited since classical times until their first documented hailing by Mediterranean sailors around the year 1336. Subsequently, the Canary Islands appeared on charts, such as *Atlas de cartes marines*, and regular sailings crossed Atlantic waters to reach them. An Atlantic slave trade began. Native peoples of the Canary Islands, known as the Guanche, were captured, taken to Mediterranean ports such as Genoa, and sold. Religious missions were undertaken, as were attempts to establish colonies. The Canary Islands eventually became the possession of the Iberian kingdom of Castile, by the Treaty of Alcáçovas in 1479.[18]

Other Atlantic islands also appear on *Atlas de cartes marines*. Six islands in the Atlantic, off the coast of Portugal, correspond to the Azores. In the fifteenth century, the Portuguese crown claimed the Azores, and the oldest surviving document citing this claim dates from 1439. The names given to the islands indicate that the chartmaker's information must have come from sailors from Italian ports who already knew of the islands before the Portuguese colonization efforts. According to the earliest document, signed by the king in 1439, Dom Henrique, known in English as Prince Henry the Navigator, received the right to plant a colony. As an incentive, the first colonists were exempted initially from any taxes. Dom Henrique introduced sheep, which would have had a major ecological impact on the islands.[19]

Also located were three islands north of the Canary Islands, which correspond to Madeira. The fact that they were already named—Porto Sto (Porto Santo), Insula de Legname, and Insula deserte—reveals that these islands were also known to sailors and chartmakers before their official "discovery" by Portuguese sailors early in the fifteenth century. In 1419, two Portuguese noblemen claimed Porto Santo, in the archipelago of Madeira, for their king. The Portuguese king made the colonization of these uninhabited islands his personal project, and the cultivation of sugar became a primary focus. By 1455 the first mill was operating.[20]

The illustrations and shorter legends placed inside the coast of West Africa on the *Atlas de cartes marines* document the well-established overland trade routes between the Mediterranean and West Africa. This longstanding trans-Saharan trade had supplied the West African kingdoms with Mediterranean trading goods, such as horses, metals, and textiles, in exchange for gold. The wealth to be gained from this trade is symbolized by the richly painted illustration of King Mansa Musa holding a nugget of gold. The chartmaker locates the starting point of the trans-Saharan route in North Africa, describes the trade in a legend, and includes various places along the route, such as Timbuktu, Gao, and Niamey. However, the *Atlas de cartes marines* also shows that competing with this well-established overland trade is the boat of Jaime Ferrer, heading for the River of Gold.[21]

The *Atlas de cartes marines* conveys new and old knowledge about places in the Atlantic accessible by mariners. Visible on it are three major Atlantic archipelagoes— the Canary Islands, Madeira, and the Azores—which would become part of the Atlantic known to mariners in the fifteenth century. The chartmaker's careful attention to the west coast of Africa, and especially the illustration of Jaime Ferrer's boat, suggests that Mediterranean merchants were already seeking maritime routes to tap into the lucrative trans-Saharan trade. This commerce would grow in the fifteenth century as boats leaving Mediterranean and Atlantic ports (such as Lisbon) loaded many of the same products traditionally carried overland. The gold known to be controlled by West African kings was a potent draw for merchants.

Yet in spite of its larger size and the detailed information provided, in the context of the whole world the Atlantic still remains peripheral on the *Atlas de cartes marines*—appearing at the far western edge of the world (figure 1.3). As compared to the Mediterranean, which had been sailed for centuries, the Atlantic had come to be known only recently and was not yet an obvious or coherent unit.[22]

Mappaemundi began to change in the fifteenth century. Woodward refers to this as a "transitional period" between the Middle Ages and the Renaissance—a time when *mappaemundi* incorporated information from charts as well as the new influence of Ptolemaic ideas. A Byzantine copy of Ptolemy's *Geography* that consisted of the text and tables but probably not the maps appeared in Italy at the end of the fourteenth century. From this, and possibly a second Greek manuscript, a clerk at the Vatican created a Latin version. As humanists copied and printed this text, it not only spawned a deep interest in mapping among Renaissance humanists but also challenged the traditional religious themes and underlying structures of the medieval T-O *mappaemundi*.[23]

A transitional *mappamundi* that shows the influence of the chartmaking tradition is the *Mappamondo Catalano estense*, also known as the *Carta Catalana*, created by an unknown chartmaker sometime between 1450 and 1460 (plate 1). It presents a slightly larger Atlantic with many of the same features as depicted on the *Atlas de cartes marines*, even as the anonymous chartmaker used a circular form. Similar to earlier *mappaemundi*, the *Mappamondo Catalano estense* has a circular frame and the known world is surrounded by oceans. Jerusalem appears near its center, and the three continents of Europe, Africa, and Asia are separated but only in part by the traditional T of the T-O. The Mediterranean separates Europe from Africa, but Asia is less separated from Africa by the Nile than by the Red Sea, the Persian Gulf, and the Indian Ocean. There is no clear separation between Europe and Asia. As on earlier *mappaemundi*, the earthly paradise is located in a specific place: East Africa. Similarly, the chartmaker used biblical, classical, and medieval sources for the legends, written in Catalan, to explain the nature of distant lands and their peoples. Richly painted vignettes of kings, towns, and animals on all three continents recall the kinds of illuminations found on T-O *mappaemundi*, and it was certainly a luxury item, likely intended for a prince or a wealthy merchant. Despite these similarities to T-O *mappaemundi*, the anonymous chartmaker created the *Mappamondo Catalano estense* using the techniques found on charts. North appears at the top, and a web of lines—the wind directions—radiate out across the land and sea. The anonymous chartmaker differentiated between known and unknown coastlines by carefully lettering in the place names at right angles on the land side of known places.[24]

Fra Mauro, a monk in the Camaldulian Order, living on the island of Murano, next to Venice, created another transitional *mappamundi* in 1450, today known *as Il Mappamondo di Fra Mauro* (plate 2). Mauro had read Ptolemy, as can be seen in the legends where Mauro frequently cites him. However, Fra Mauro rejected many of Ptolemy's mapping tools because they decentered Jerusalem. Instead, Fra Mauro created a very distinctive *mappamundi* into which he compiled all the knowledge of the world he could collect, including place names, visual imagery, and written descriptions. Mauro's sources included the narratives of Marco Polo, Coptic Christian and Muslim informants from East Africa, Arabic sources, and charts. Fra Mauro retained a T-O structure but positioned south at the top. He located the Garden of Eden outside his map of the known world, placing it in the lower left border.[25]

On these two transitional *mappaemundi*, the Atlantic appears in the same place—on the western edge of the world—but it has more space on the *Mappamondo Catalano estense*. In figure 1.4, the two *mappaemundi* are redrawn and juxtaposed. On each, the Atlantic is part of the oceans that surround the known world, but its greater size and extent on *Mappamondo Catalano estense* are noticeable.

The geographic detail provided for the Atlantic on each of these transitional *mappaemundi* incorporates both traditional and contemporaneous knowledge. On the *Mappamondo Catalano estense*, the chartmaker used a traditional name for the Atlantic: "Mar Ocleana." Immediately to the south of the Canary Islands, Fra

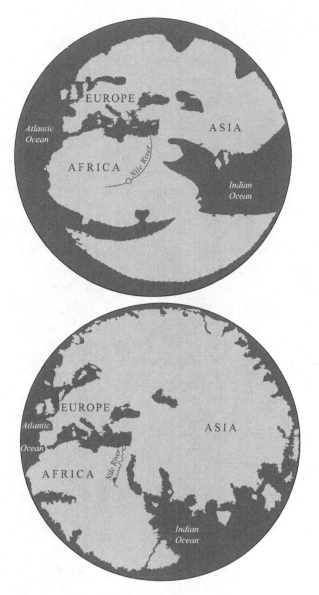

Fig. 1.4. The Atlantic on Transitional *Mappaemundi.* Redrawn from *Mappamondo Catalano estense,* ca. 1450–60 (*top*), and *Il mappamondo di Fra Mauro,* 1450 (*bottom*).

Mauro provides the name: "Oceanus Athlanticus" (Atlantic Ocean). As on the *Atlas de cartes marines,* the Atlantic archipelagoes—the Azores, Madeira, and the Canaries—appear on both. An example of new knowledge is the inclusion on both world maps of the island named Arguim, the site of an early trading post established in West Africa by the Portuguese prince, Dom Henrique.

The island of Arguim is not emphasized by the anonymous chartmaker of the *Mappamondo Catalano estense* but it is named, and a boat sailing for the West African coast is heading to it (or very near to it). Alvise Ca' da Mosto (also spelled Cadamosto), a Venetian who made two voyages to West Africa (1455–56), visited Arguim. He described Arguim as an island in a gulf with very good fish. However, he remarked that the waters were dangerous and care must be taken to sail only during the day and with the sounding line always in hand. On the island, Ca' da Mosto writes that Dom Henrique had established a ten-year monopoly, which he hoped to make permanent, that allowed only boats under his license to trade there. Arguim is an example of a *feitoria* (fortified trading post), built by Dom Henrique in his quest to monopolize the trade with West Africa. Arguim was constructed in 1448; and by the 1450s, as Ca' da Mosto observed, a regular commerce in textiles and grains exchanged for slaves and gold was taking place.[26]

Below the boat headed for Arguim on *Mappamondo Catalano estense* are two stylized islands—one red, the other blue—named Cades, with a legend stating "Cades Island here Hercules placed two columns." The Pillars of Hercules have moved, as it were, to mark the current limits of the known Atlantic. South of Cades, the chartmaker provides no more detailed information, implying that beyond lies the unknown ocean as well as the unexplored lands of Africa. In figure 1.5, the western periphery of the *Mappamondo Catalano estense* is redrawn in order to illustrate how much of the Atlantic the chartmaker included, even as it still remained peripheral in the world. The extent of the coastline to Cape Verde is carefully mapped, and the location of the sailboat reveals that the chartmaker strategically placed it within the range of voyages of the time to West Africa.[27]

Ca' da Mosto described sighting the Cape Verde Islands on his second voyage (1455–56) to the west coast of Africa. He and his co-voyagers had just passed the Canary Islands and were within sight of Cabo Branco, which was just to the north of Arguim and on their route, when a violent windstorm blew them out to sea. Three days later they saw islands, previously unknown, which were part of the Cape Verde Archipelago. The Cape Verde Islands do not appear on the *Mappamondo Catalano estense* nor does Fra Mauro include them on his *Mappamondo*. Both do, however, include Cape Verde as a prominent promontory on the West African coast. On both *mappaemundi*, it marks the end of their detailed knowledge about West Africa.[28]

Fra Mauro's Atlantic is smaller than on the *Mappamondo Catalano estense* and less easily understood visually; nevertheless, he includes a great deal of detail, indicating his careful study of the region. On the map is a feature he labels the "Sinus Ethyopicus" (Gulf of Ethiopia) just south of Cape Verde. Fra Mauro places the label "Regno museneli" along the coastline of the Sinus Ethyopicus, and the label is,

Fig. 1.5. Atlantic Details on a Catalan World Chart Redrawn from *Mappamondo Catalano estense*, ca. 1450–60.

according to historian Piero Falcetta, a reference to the kingdom of Mali. Right offshore is a sailing vessel. Across the bay, south of where he locates Cape Verde, Fra Mauro writes a legend that reflects his interpretation of the two stylized islands, labeled Cades on the *Mappamondo Catalano estense*. Fra Mauro writes: "I have often heard many say that here there is a column with a hand and inscription that informs one that one cannot go beyond this point. But here I would like the Portuguese that sail this sea to say if what I have heard is true, because I am not so bold as to affirm it."[29] Seeming to doubt that there is any such column, Fra Mauro instead places the Straits of Hercules immediately beyond the Straits of Gibraltar. On the map, the text "Fretum herculeum" (Straits of Hercules) appears in waters he names "Mar Gaditaneum" (the Gaditanean Sea). Farther out in the ocean, next to where he has placed the island of Madeira, he makes this comment: "Note that the Columns of Hercules refer to nothing but the division of mountains which once, according to the fable, closed the Strait of Gibraltar."[30]

Fra Mauro followed a common stylistic pattern of using curving blue lines to represent water, but he carried this artistic rendition one step further by transforming the generic wavy lines into groups that suggest currents. These currents move through the Straits of Gibraltar and change directions in the Atlantic. Sailboats follow currents north to ply the waters of Iberia, France, England, Ireland, and the North Sea. To the south, similarly, the currents carry boats to ports of call along the coast of West Africa.

Fra Mauro carefully considered Ptolemy but ultimately chose not to create a *mappamundi* using one of Ptolemy's projections. Ptolemy's ideas about geography did, however, greatly influence other Renaissance humanists, mapmakers, and artists. A Greek scholar, Manuel Chrysoloras, carried the Greek text of Ptolemy's *Geography,* and perhaps twenty-seven maps redrawn as described in the text, from Constantinople to Florence in 1397. Welcomed by a group of scholars eager to learn Greek, Chrysoloras began translating the text into Latin, an undertaking that was finished in Rome by the Vatican clerk, Jacopo d'Angelo da Scarperia between 1406 and 1409. In the translation, Ptolemy's *Geōgraphikē* became *Geographia* or *Cosmographia* in Latin (today this work is known as Ptolemy's *Geography*). The humanist Greek scholars in Florence re-created the maps, and copies of Ptolemy's *Geography*—with text, maps, and tables—quickly became desirable items for Renaissance humanists to own.[31]

Ptolemy saw the world within a larger universe, and he conceptualized the earth in relation to the stars and the sun. He was interested in the circumference of the earth and in the distances between locations of various places on earth. These, he believed, could be known through empirical observation of stars and the application of mathematical principles. Ptolemy's other works, on astronomy and astrology, reinforced his approach to geography among humanists. On land, the humanist scholar could observe stars and make calculations that created knowledge about places in the world. Ptolemy showed not only the implicit harmony in the universe but also how one could add to knowledge through careful observation and study.[32]

The eight sections of Ptolemy's *Geography* became eight volumes in the editions that appeared in Renaissance Europe. Volume 1 begins with a definition of world cartography: "an imitation through drawing of the entire known part of the world together with the things that are, broadly speaking, connected to it." The purpose, or "essence" of world cartography, Ptolemy states, is "to show the known world as a single and continuous entity, its nature and how it is situated," which takes into account only the most important things, "such as gulfs, great cities, the more notable peoples and rivers." According to Ptolemy, cartography requires systematic research, careful attention to recent findings, and a critique of what has been done before. Ptolemy establishes the importance of latitude "from north to south" and longitude "from east to west" to calculate the location of places. He defines what must go into a world map and addresses the problem of projection—how to represent the spherical earth on a flat plane. Ptolemy thus provides a clear definition of what cartography is, presents a model for mapmakers to draw their maps on flat planes, and delivers a clear means to locate places—identified by the intersection of their lines of latitude and longitude. Volumes 2 through 7 consist of tables of latitude and longitude for sites in Europe, Africa, and Asia. Volume 8 describes how to make the regional maps that Ptolemy deems necessary: twelve maps for Asia, four for Africa, and ten for Europe.[33]

Editions of Ptolemy's *Geography* contained a *mappamundi*, and before 1500 the Atlantic is not mapped with the detail visible on either the *Atlas de cartes marines* or the two transitional *mappaemundi*—the *Mappamondo Catalano estense* and *Il Mappamondo di Fra Mauro* discussed previously. The Atlantic remains peripheral. This can be seen in the richly illuminated version of Ptolemy's *Geography* acquired by Borso d'Este, the Duke of Ferrara in 1466. Nicolò Germano prepared the manuscript using the Latin text by Jacopo d'Angelo da Scarperia. The text is written by hand in handsome italics—the script favored by humanists—and elegant illuminations of capital letters begin each section. The last volume is devoted to the twenty-seven maps: each map is spread out over two facing folios, with the bifolio measuring 45 × 62 cm, although the maps themselves are smaller.[34]

The first map is an arresting world map, known today as *Planisfero*, that Nicolò Germano created using Ptolemaic principles (plate 3). Having carefully copied Ptolemy's text, Nicolò Germano knew that, according to Ptolemy, the goal of a world map was to "show the known world as a single and continuous entity." By this Ptolemy meant that only the most important features should be stressed, such as great cities, notable peoples, rivers, gulfs, and broad general outlines. Although Ptolemy provided an analogy (he states that the world map is like a portrait of the head), he did not see the mapmaker as an artist. Rather, the mapmaker needed to be familiar with mathematics.[35]

The first step for Nicolò Germano to create a world map was to lay out the projection. Ptolemy provided explicit instructions, beginning with a "surface in the shape of a rectangular parallelogram" followed with steps that used basic geometry. Nicolò Germano's adoption of Ptolemy's first projection would shape how the Atlantic would appear within the world. The most radical change, however, was not the mapping of the Atlantic; rather, it was the new location of Jerusalem, which would no longer appear at the map's center.[36]

Once the projection was set out and the geographic information entered, Nicolò Germano made some design choices. He represented Europe, Africa, and Asia as ivory-colored land masses cut by blue rivers and broken up by long rolling brown ropelike mountain ranges. He selected a deep blue for the oceans.

The "Oceanus Occidentalis" (Western Ocean) appears at the far west, and it extends south along the coast of Africa, and north to the British Isles and the North Sea. This Western Ocean is smaller and less consequential than the massive "Mare Indicum" (Indian Ocean), which is presented as Ptolemy envisioned it: a landlocked sea in Asia. Beyond the inhabited world and the known oceans, and outside of the projection with its lines of latitude and longitude, Nicolò Germano paints a huge expanse of blue. This is the celestial sphere, with the sun at the top, clouds interspersed in the blue, and twelve windheads carefully painted in white to make them visible against the sky. Each head is named to represent the major characteristics, and direction, of its wind.

The importance given to the celestial sphere, which appears as a kind of border to the *mappamundi*, does not completely diminish the space that the Atlantic Ocean can occupy. Ptolemy's projection did allow the Atlantic Ocean to be mapped as far north as England, Scotland, and Ireland, and along the coast of West Africa in the south, as can be seen in the redrawing of this map in figure 1.6.

Ptolemy's writings opened up exciting new ways for Renaissance artists and cosmographers to explore world geography and to make beautiful maps. However, Ptolemy's information—such as the specific coordinates of latitude and longitude he presented for mapping various locations—was all very old. New knowledge, especially of the Atlantic Ocean, was difficult to place using Ptolemy's instructions. The rich detail visible for the Atlantic Ocean on the *Atlas de cartes marines,* as well as on the transitional *mappaemundi* created by Fra Mauro

Fig. 1.6. The Atlantic on a Ptolemaic *Mappamundi.* Redrawn from Nicolò Germano, *Planisfero,* in Ptolemy, *Cosmographia,* 1466.

and the anonymous chartmaker of the *Mappamondo Catalano estense*, does not appear on Nicolò Germano's *Planisfero*, even though all of these maps and charts had been created earlier.

Henricus Martellus Germanus, a German artist and mapmaker living in Florence in the second half of the fifteenth century, was one of the first mapmakers to incorporate detailed information on the Atlantic available on charts into a Ptolemaic *mappamundi*. Skilled as an artist and mapmaker, Martellus created illuminated manuscripts, such as copies of Ptolemy's *Geography*, for prominent patrons. He also created sumptuous manuscripts based on a book of the Aegean islands, and in one of these copies Martellus greatly modified the standard Ptolemaic world map. The manuscript known as the *Insularium Illustratum* consisted of maps of islands, and it was based on an original made by the Florentine priest Cristoforo Buondelmonti, who had visited many of the Greek islands and had created an illustrated itinerary of his travels in four separate editions between 1418 and 1430. Originally copying the manuscript closely, but later adding new maps and changing the text, Martellus recreated a charming book, most likely on commission, for prominent patrons. In several of the surviving versions, Martellus included a *mappamundi*, which did not appear in Buondelmonti's original, or even in Martellus's first versions.[37]

All the *mappaemundi* by Martellus in copies of the *Insularium Illustratum* diverge from traditional Ptolemaic representations of Africa and the South Atlantic. The most divergent is the *mappamundi* (measuring 340 × 240 mm) that appears on pages 68 verso and 69 recto of the British Library edition of Martellus's *Insularium Illustratum* (figure 1.7). On this *mappamundi*, Martellus redrew the coastline of Africa in order to take into account information known in Lisbon following the return of Bartolomeu Dias in 1488. Bartolomeu Dias had left Lisbon in 1487, sailed along the entire coast of West Africa, rounded the Cape of Good Hope, and reached Rio do Infante (today Great Fish River), before returning to Lisbon in December 1488. Apparently, Martellus had access to these charts, or copies of them, for Martellus maps the western coastline of Africa, from Gibraltar to the Cape of Good Hope, and he shows the beginnings of the Indian Ocean coastline of southeast Africa.[38]

The coastline of Africa breaks the frame of Martellus's *mappamundi*. In this edition of the *Insularium Illustratum* Martellus frames all of his maps with similar borders, and he occasionally does break the frame in interesting ways. Therefore, there is nothing unusual in his choice to break the frame, but the broken frame does draw the viewer's attention to the startlingly new connection between the Atlantic and Indian Oceans. Normally, a map viewer familiar with Ptolemaic *mappaemundi* would expect to see a landlocked Indian Ocean,

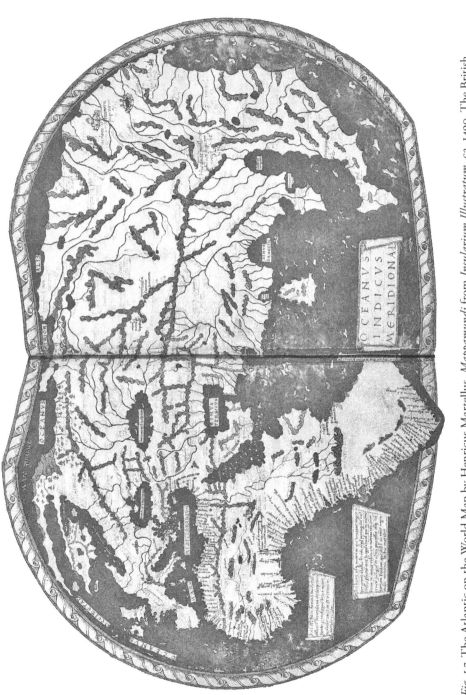

Fig. 1.7. The Atlantic on the World Map by Henricus Marellus. *Mappamundi* from *Insularium Illustratum*, ca. 1490, The British Library, London.

but on Martellus's *mappamundi* there is a clear passage from the Atlantic to the Indian Ocean.

Martellus's *mappamundi* includes detailed new information on the South Atlantic that Martellus supplies through textual legends, images, and place names. This information not only documents Dias's voyage but also those before, such as that by Diogo Cão, who made two trips to Africa, between 1482 and 1486, crossing the equator both times. On his expedition to Kongo and Angola, Cão erected stone markers (*padrões*) at the mouth of the Zaire River, Cabo Lobo, Capo Negro, and Cabo do Padrão (Cape Cross). Martellus enters place names in Portuguese, some of which explicitly mention the padrões, such as "Ponta da padron" (Marker Point).[39]

Martellus places two major legends, written in Latin, in the South Atlantic, and they are so prominent that an attentive map viewer would not miss them. The first makes a clear statement about the geography of the African continent. Martellus writes: "This is the modern true shape of Africa according to the map of the Portuguese, from the Mediterranean Sea to the southern ocean."[40]

Martellus's subsequent huge wall map of circa 1491, now at Yale University where it is called *Map of the World of Christopher Columbus*, similarly reflects Martellus's interest in Portuguese voyages along the coast of Africa south of the equator—as well as his willingness to break from Ptolemy. Brilliantly studied by historian Chet Van Duzer through the use of multispectral imaging, this later *mappamundi* is freestanding and very large, measuring 112 × 201 cm (approximately 4 × 6½ feet). Martellus used a modified form of Ptolemy's second projection, the pseudo-cordiform or heart-shaped projection. The Atlantic is again connected to the Indian Ocean, and the continent of Africa is massive, with an oddish shape given to the southernmost coast. The technique of multispectral imaging allows Van Duzer to read the formerly illegible texts on the map, and from them it is clear that this *mappamundi* proved highly influential to later mapmakers. The place names entered along the African coast south of the equator confirm Martellus's knowledge of, and high regard for, the Portuguese charting of Africa following the voyages of Diogo Cão and Bartolomeu Dias. But the information on the map predates, Van Duzer concludes, the 1497–98 voyage of Vasco da Gama that reached India.[41]

Artists and mapmakers also made globes, which provided another way to visualize the Atlantic Ocean. Because they are round, globes convey the geography of the world more intuitively than maps and charts, which, created on flat surfaces, inevitably distort geographic information. As a result of their spherical form, the Atlantic appeared on globes as an ocean bounded on both sides by continents. The form of the globe, therefore, made the Atlantic less peripheral, and

because the globe could be turned and viewed from different angles, the whole
Atlantic could be seen.

Globe making dates back to classical Greece, and it became popular in Europe
during the Renaissance. The technique for making a globe first required making
a map on paper or parchment, but in such a way that the surface could be cut and
pasted around a ball. The globe could then be painted, making it an attractive
object. Martin Behaim, a merchant from Nuremberg who lived for some time in
Lisbon, commissioned a globe from artisans in Nuremberg around the year 1492.
Behaim supplied the information, and the artisans created the globe and painted
it. Behaim's sources were similar to those of any other humanist cosmographer:
travel narratives, Ptolemy's *Geography*, classical texts, and maps he consulted. One
of his sources was likely a large, elaborate, printed map similar to Martellus's large
mappamundi of circa 1491. The Behaim-Globus, which survives to this day, may
be seen in the German National Museum in Nuremberg. As described by Ernest
George Ravenstein, the globe has a circumference of 507 millimeters (20 inches)
and is "crowded with over 1,100 place names and numerous legends in black, red,
gold, or silver." In addition, there were 111 miniatures, which included crests, flags,

Fig. 1.8. The Atlantic on Behaim's Globe. Created from a redrawing of Behaim's Globe
(1492) in Paullin and Wright, *Atlas of the Historical Geography.*

coats of arms, kings seated on thrones or inside tents, portraits of saints (Peter, Paul, Mattthew, and James), missionaries, travelers, boats on the seas, fish, animals, and a few monsters—Sciapodes, for example, in South Central Africa. Behaim's Atlantic is arresting because of the fact that it is fully incorporated into the globe. On flat maps, much of the Atlantic was omitted and left to the imagination. However, in Behaim's case, the globe included the entirety of the extent of what he believed to be the Atlantic Ocean between Europe and Africa and Asia. Behaim fills the Atlantic Ocean with islands—all of the previously mapped archipelagoes appear, as do ships and sea creatures. He bisects the ocean with geographic parallels and meridians, such as a graduated meridian located west of the Azores, the equator, and the Tropics of Cancer and Capricorn. Many textual legends occupy the space. Behaim's globe implied that although the Atlantic remained deeply unknown beyond the coastlines of Western Europe and Africa, it was fully accessible to mariners. In figure 1.8, Behaim's globe is re-created from a line drawing of the globe gores done early in the twentieth century. This reconstruction reveals how the Atlantic could be more easily conceptualized on a globe than on a flat map. Nevertheless, even on a globe, the Atlantic is the back side of the world.[42]

From the time of Anaximander's lost maps in ancient Greece to 1500, distinct mapping genres in the West visualized the Atlantic. Globes gave the greatest coherence to the Atlantic, followed by charts. Parts of the Atlantic Ocean were known and familiar, and knowledge of the ocean's waters tended to be concentrated in maritime communities. As chartmakers added locations, such as islands and coastlines, they expanded the mapping of the Atlantic Ocean. Yet even as chartmakers supplied new knowledge and even as mapmakers began to incorporate it on their world maps, classical and medieval preconceptions continued to shape the understanding of the Atlantic Ocean as a peripheral space. This perception, however, would change dramatically after 1500.

CHAPTER TWO

The Year 1500

Although 1492 traditionally marks the discovery of the continents in the western Atlantic, and one would therefore expect to see the Atlantic Ocean appearing immediately as large and expansive on world maps, no surviving historical maps show this until 1500. The delay is notable, for Columbus was known to have been a chartmaker, and he promised Ferdinand and Isabella "a new navigational chart in which I will place the whole ocean and all of its lands, in their proper places." Neither this nor any other world charts by Columbus have endured; only his quick sketch of the northern coast of Hispaniola survives. In 1500 Juan de la Cosa, who sailed with Columbus on his first two voyages of 1492 and 1493, created the oldest existing world chart, known today as the *Carta Universal*, after returning from his third transatlantic crossing. This chart showed extensive lands in the western Atlantic Ocean. Two years later, a world chart created in Lisbon, and known as *Carta del Cantino*, presented a striking new vision of the world in two hemispheres—the western centered on the Atlantic Ocean and the eastern on the Indian Ocean. By 1506 a larger and more detailed Atlantic Ocean appeared on a printed map, and in 1507 Martin Waldseemüller gave the southern lands of the Western Atlantic the name "America." These charts and maps are the earliest surviving cartographic documents that depict the Atlantic Ocean bordered by continents in the East and the West. They conveyed a dramatic new idea of the size, extent, and importance of the Atlantic Ocean in the world.[1]

The event of new discoveries was not enough to change mapping conventions immediately. Columbus's promised "new navigational chart" would have shown not a new view of the Atlantic Ocean but only a somewhat larger one. When Columbus made his first exploratory voyage west in 1492, he believed he had reached India. As historian Nicolás Wey Gómez explains, Columbus and his European contemporaries understood India to be "a vast geographical system" that extended beyond what we know today to be the Bay of Bengal to include Indochina, Malaysia, the South China Sea, and Indonesia. Columbus expected to encounter islands, because Marco Polo had reported archipelagoes, but he anticipated making landfall on mainland Asia. Therefore, his view of the ocean he crossed (and how it should be mapped) fit into the prevailing understanding of the Indian Ocean, as depicted on Ptolemaic *mappaemundi* (see plate 3) and other late fifteenth-century maps, such as Martellus's huge *mappamundi* of circa 1491 or Behaim's globe of 1492 (see figure 1.8). Any maps or charts made by Columbus following the first voyage in 1492 would have placed the Bahamas, Hispaniola, and Cuba in the South China Sea. On his second voyage in 1493, when Columbus commanded seventeen ships, he discovered more islands, including Puerto Rico and Jamaica, and he further explored the islands of Hispaniola and Cuba, which he first visited in 1492. He still believed that they were sailing in the expansive Indian Ocean and that the long southern coast of Cuba was mainland China. In June 1494, when three caravels under his command had almost reached the westernmost point of the southern coast of Cuba, Columbus required those sailing on the *Niña*, *San Juan*, and *Cardera* to swear that the island of Cuba was a mainland, on pain of being fined (or, in the case of a deckhand, whipped) and having one's tongue cut out. The chartmaker Juan de la Cosa took the oath, as did the others, that "he had never heard nor seen an island with a coastline of three hundred and thirty five leagues from west to east . . . and that he had no doubt but that it was mainland." On his third voyage in 1498, Columbus reached the coast of northern South America, discovering "this our great mainland," according to Bartolomé de las Casas (the sixteenth-century Dominican missionary priest and historian). The great mainland was South America. But on his fourth and last voyage in 1502, Columbus sailed along the eastern coast of Central America, still seeking a passage into the Indian Ocean.[2]

Juan de la Cosa, who signed the oath that Cuba was part of Asia, created the *Carta Universal*—the first world chart to show an expansive Atlantic and a massive continental presence in the West—soon after returning from his third transatlantic voyage. According to Bartolomé de las Casas, La Cosa was "the best pilot of those waters, because he was on all of the voyages made by the Admiral [Columbus]," and the Spanish court chronicler Peter Martyr d'Anghiera called him a "man of such great influence that he was named by royal commission to be royal master of ships." Moreover, Anghiera describes standing in the room of Bishop Juan Rodríguez de Fonseca (who administered the affairs of the Spanish Indies) and holding in his hands what he called "a globe with these discoveries" and many "parchments that the sailors call navigational charts," and he noted that the most commendable of all were those by Juan de la Cosa. Scholars speculate that the world chart of 1500 was intended for Fonseca or for the monarchs Ferdinand and Isabella. Even if, as some scholars have suggested, the world chart dated and signed by la Cosa in 1500 may in fact have been made by a group of chartmakers who combined their information, La Cosa possessed his own extensive, firsthand knowledge of the Spanish discoveries, then known as the Spanish

Antilles or the Spanish Indies. The world chart represents his or his group's synthesis of the location of these places in the world in 1500.[3]

La Cosa's *Carta Universal* measures 93 × 183 cm (about 3 × 6 feet) and is painted with colored inks over a surface fashioned from two joined calfskins (plate 4). According to Joaquim Alves Gaspar, the extent of the world depicted on the world chart encompasses approximately 240 degrees, or two-thirds of the earth's surface, and La Cosa used Mediterranean charts for Europe and the Mediterranean Sea and the Black Sea, Portuguese charts for the coast of West Africa, and Ptolemaic maps for the Indian Ocean. The information for the Americas came from sketches—presumably his own—and pilots' descriptions from Spanish and English voyages prior to 1500.[4]

La Cosa presents the Atlantic Ocean as a large body of water, and he gives it a generic Latin name: Mare Oceanum (Ocean Sea). An unnamed but enormous landmass (North and South America), painted in deep greens, reaches around the Mare Oceanum. As compared to the mainland, La Cosa leaves the islands of Hispaniola and Cuba unpainted so that their coastal place names may be easily read. Mapped in this way, the Spanish Indies appear known, just as are lands on the other side of the ocean—in Africa, Europe, and Asia. A few place names are written out opposite the densely green coastline of northern South America, and Spanish flags mark places claimed by Spain. Farther south, near the equator, toponyms (place names) appear, while flags and ships are painted off the coast; this was the region visited by La Cosa and Alonso de Ojeda on their 1499 voyage, as well as by Vicente Yáñez Pinzón in the same year.

La Cosa, or an artist working with him, painted an icon of St. Christopher carrying the Christ Child over the isthmus of Central America. The image appears almost as an open window or a door, perhaps suggesting that through it lay the way to the Indian Ocean. In figure 2.1, which shows this detail from La Cosa's *Carta Universal*, St. Christopher is on land, but the waters beyond suggest possible new destinations to the west. Below the image are the islands of Cuba and Hispaniola, as well as others in the Spanish Indies. Under the icon appears the text: "Juan de la Cosa made it [this chart] in Puerto de Santa Maria in the year 1500."[5]

On *Carta Universal*, La Cosa maps the connection between the Atlantic and the Indian Oceans in the East. He records in two short legends the history of the two key Portuguese voyages along the coasts of Africa that had proved this fact: the first voyage, led by Bartolomeu Dias, who rounded the Cape of Good Hope in 1487–88, and the second, led by Vasco da Gama, who sailed north along the east coast of Africa and across the Arabian Sea to reach the Malabar Coast of

Fig. 2.1. St. Christopher. Detail from Juan de la Cosa, *Carta Universal,* 1500, Museo Naval de Madrid & Biblioteca Virtual del Ministerio de Defensa, Madrid.

India in 1498. The legend referring to Dias, "The excellent King Don Juan [King João II of Portugal] discovered to here," is placed on the southeast coast of Africa. Near India, the legend pertaining to Vasco da Gama reads: "Land discovered by King Manuel of Portugal." La Cosa recorded toponyms for the entire coast of Africa as well as the west coast of India. Illustrations of ships flying Portuguese colors sail along the west coast of Africa and round the Cape of Good Hope and are gathered off India.[6]

As La Cosa was working on his world chart in Santa Maria, not far away in the Spanish city of Seville Amerigo Vespucci began a letter to his patron, the powerful Lorenzo di Pierfrancesco de' Medici, of Florence. Vespucci had sailed on the very same voyage with La Cosa and Alonso de Ojeda, although it is unclear whether he sailed as an adventurer, merchant, or pilot. After the formal salutation, Vespucci begins, "The present letter is to give you the news . . . because I believe Your Magnificence will be pleased to hear of all that took place on the voyage, and of the most marvelous things I encountered upon it." Historians know that Ojeda led the first voyage to break the monopoly held by Columbus and that it consisted of four boats that departed from the port of Cadiz in May 1499. After provisioning in the Canary Islands, the four vessels crossed the Atlantic Ocean and landed on the coast of northern Spanish America. Then the expedition apparently split, as one part—with Vespucci—sailed south, across the equator, entering the Amazon River basin but eventually turning north for present-day Suriname, the Guayanas, and Venezuela. The expedition visited the islands of Trinidad, Curaçao, Margarita, and Hispaniola before sailing back to Spain with a cargo of two hundred Indian slaves.[7]

Vespucci thought his letter important, and he knew it would be long. "And as I am somewhat given to prolixity," he writes, "it might best be taken up and read when you have amplest space for it." Near the end of the letter, Vespucci makes a promise: "Just as I have given you an account by letter of what happened to me, I shall send you two depictions of the world, made and ordered by my own hand and knowledge: one chart will be a flat rendering and the other a map of the world in spherical form." Vespucci sounds confident that he could produce the map and globe he promised to his patron, and he already had someone in mind who could deliver them to Florence. However, if Vespucci in fact made this map and a globe, there is no confirmation that they ever arrived in Florence.[8]

If such a map and globe were created, how would they have presented the Atlantic Ocean and its relation to the larger world? Peter Martyr d'Anghiera believed that Vespucci had assisted in the creation of a Portuguese map held by the Bishop Fonseca and noted also that Vespucci "was very skilled in this art." Yet, beginning with Las Casas, scholars have not particularly trusted Vespucci. Las Casas saw Vespucci as "eloquent" but also as one who dishonestly took credit for things that others did. Vespucci's modern biographer Felipe Fernández-Armesto calls him a "pimp in his youth and a magus in his maturity," hardly a compliment. Nevertheless, Vespucci's promise of a map and a globe made by his own hand opens up reflections on how the early maps that included the extent of the Atlantic Ocean might have been made and for whom, and how they challenged existing mapping paradigms. Even in the absence of actual maps, charts, and globes, the references to them in letters—such as Vespucci's—help to reconstruct how the Atlantic Ocean was taking shape.[9]

A close reading of Vespucci's letter reveals still more about the map and globe he intended to send. Vespucci states to de' Medici that he had already made a globe for Ferdinand and Isabella, adding that "they prize it highly." This suggests that it was a carefully crafted, aesthetically pleasing object to be admired. On the map for his patron in Florence, Vespucci planned to include the coast of Africa in order to show the way to India, for he directs his patron's attention to the map where he (Vespucci) will have recorded information from the first Portuguese voyage to return from India. Addressing his patron, Vespucci writes, "as you will see by the map, they [the Portuguese] navigate continually in sight of land, and sail along the entire southern part of Africa." Vespucci reveals that one of his reasons for joining Ojeda's expedition was "to see whether I could round a cape of land which Ptolemy calls the Cape of Cattigara, which is near the Sinus Magnus [Great Bay]." As these refer to places in Asia that were commonly placed on Ptolemaic maps, Vespucci reveals that he thought he was headed to the Indian Ocean. The Sinus Magnus, which corresponds to the modern Bay of Bengal, appears on the *Tabula Undecima Asie* (Eleventh Map of Asia) in Ptolemy's *Geography*. Thus, on his return, Vespucci, like Columbus, believed he had reached the edge of the Indian Ocean.[10]

Vespucci describes their position, as they sailed along the coast of South America, as within the "Torrid Zone" well south of the Tropic of Cancer, and he relates that when they reached the equator, they "completely lost sight of the North Star." From the way he phrases these passages, it seems that Vespucci envisioned his patron reading his letter while studying the map and globe he would provide. Vespucci also suggests his patron might consult with learned persons in Florence who understood "la figura del mondo" (the shape of the world). Some of these learned persons in Florence, Vespucci adds, "perhaps may wish to emend something in what I have done; nonetheless let him who would emend me wait for my coming, as it may be that I shall defend myself." From this we can ascertain that Vespucci thought his map and globe would be circulated among intellectuals and perhaps be controversial. Seven years later, Queen Juana of Castile appointed Amerigo Vespucci as *piloto mayor* (head pilot) and commissioned him to make the *padrón real* (royal pattern chart) in 1508. As *piloto mayor* in Spain, Vespucci taught and tested navigators, and he had the responsibility for creating and maintaining the *padrón real*. Thus, by 1508 Vespucci had acquired the reputation as the most reliable mapmaker in Spain. Therefore, it is entirely likely that he could have made a map and globe in 1500.[11]

As he brings his letter to a close, Vespucci remarks that "the King of Portugal has once more equipped twelve ships at very great expense, and has sent them to those parts [India], and surely they will accomplish great things, if they return safely." This armada, commanded by Pedro Álvares Cabral, consisted of thirteen ships and between twelve hundred and fifteen hundred men; the objective was to retrace Vasco da Gama's voyage to India. That would take Cabral deep into the South Atlantic, around the Cape of Good Hope, up the coast of East Africa, and across the Arabian Sea to Calicut. Cabral carried letters from King Manuel for the rulers of the port cities he would visit in East Africa—present day Mozambique, Tanzania, Kenya—as well as to the zamorin (the Hindu sovereign) of Calicut. On the outward journey, whether intentionally or by chance, Cabral anchored off the coast of present-day Brazil. He tarried only ten days, but before weighing anchor and setting his course for the Cape of Good Hope, Cabral sent his supply ship back to Lisbon with letters informing the king of his landing in Brazil. Vespucci apparently wrote to his patron about the return of Cabral's supply ship and the news it brought—the discovery of Brazil—but this letter has been lost. On receiving the letters sent by Cabral, King Manuel outfitted an expedition to explore the long coastline of South America. Vespucci gained passage on this voyage to Brazil, and it too became the subject of letters to his patron in Florence.[12]

In June 1501, Vespucci wrote to de' Medici from Cape Verde, where the outgoing expedition to Brazil had a chance meeting with two ships from Cabral's armada making ready for the last leg of their journey home to Lisbon. Vespucci wrote to his patron that he had "very lengthy conversations" with the men on these two ships "about the coast of the land along which they had run, and the riches they found there, and what the inhabitants have." Vespucci then provided a long summary of what he had learned, in terms of both geography and trading goods, and he entrusted the letter to a man on one of the ships heading for Lisbon.[13]

Vespucci's report from Cape Verde provides both geographic and commercial information about Cabral's voyage. He states that Cabral headed south by southwest from Cape Verde, reaching in twenty days "a land where there lived a white and naked people," a description of Cabral's landing in Brazil. Then Cabral sailed southeast by east, through a violent storm that sank several ships, and blew the remaining ships—bare-masted—past the Cape of Good Hope. Recovering from this disaster, Cabral set his course north by northeast and followed the east coast of Africa, visiting Zafale (Sofala), Mezibino (Mozambique), Chilua (Kilwa), Modaza (Mombaça), Melindi (Malindi), and Mododasco (Mogadishu). Vespucci lists names of other ports, which cannot be positively identified, but he also continues to cite places that can be identified, such as Hoden (Aden) and Ormuz (Hormuz). Vespucci states that one ship visited Meca (Mecca), and he writes that at the port of Giuda (Jeddah) "spices, drugs, and jewels" are unloaded for the caravans of camels from Cairo and Alesandria (Alexandria). He names many ports along the Persian Gulf and Indian Ocean, most of which are not recognizable, but the Indian ports of Mangalur (Mangalore), Calicut, and Cochin are. Informed by a member of the expedition named Gaspar, Vespucci learned about places that the sailors did not visit, such as Scamatara (Sumatra) and Zilan (Ceylon).[14]

Educated in Florence, and familiar with Ptolemy's *Geography*, Vespucci tried to link the information he received in Cape Verde from the sailors returning from India with what he knew from Ptolemy. When describing the ports from the Red Sea to the Persian Gulf, Vespucci writes to his patron, "I believe this to be the province which Ptolemy calls Gedrosia." Here Vespucci refers to a province in Asia that appears on Ptolemy's ninth map of Asia, and in his letter Vespucci claims to have corrected the sailor's misunderstanding with his own superior knowledge. However, in the long list of names of places cited by Vespucci in this letter to his patron, many do not come from Ptolemy but seem to have been the names that the sailors had heard and which Vespucci seems to have written down as they were spoken to him. In the letter, these toponyms appear sequentially, as they were encountered by the sailors on Cabral's voyage. This information, however, did not match up with the mental map that Vespucci carried in his head of the Indian Ocean, which was based on Ptolemy's maps as they appeared in editions of the *Geography*. After acknowledging that many of the names he reports will seem strange to his patron, Vespucci then offers an explanation: "Believe, Lorenzo, that what I have written thus far is the truth, and if the provinces, kingdoms, names of cities and islands in the ancient writers do not appear here, it is a sign that they have changed, as we see in our own Europe, where it is a great rarity to hear an ancient name."[15]

Not only was much of the nomenclature Vespucci reported unrecognizable on the maps in Ptolemy's *Geography*, but these maps did not show the Malabar Coast of India where Cabral had loaded cargoes of spices. Nor did the sailors speak of the legendary island of Taprobane, illustrated in Ptolemy's "Duodecima Asie Tabula" (twelfth map of Asia). Moreover, Vespucci believed that his voyage, which was about to head West across the Atlantic, would reach the very same places just visited by Cabral's fleet. "I hope in this voyage of mine," he writes, "to revisit and traverse much of the aforementioned area, and to discover much more."[16]

Vespucci was not alone in writing down what he had learned from sailors returning from overseas voyages. Knowledge about newly discovered lands circulated freely in the maritime communities of the port cities of Spain and Portugal, and the news spread quickly. Italian merchants, dukes, and other potentates received written reports from their ambassadors and informants on the discoveries. Especially important to Venetians was the news of the return of Cabral's armada, which threatened to break their monopoly over the spice trade. In July 1501, the remaining ships from Cabral's fleet returned to Lisbon. An emissary of the Signoria (Governing Body) of Venice wrote a letter dated 27 July 1501 recounting the details of Cabral's voyage, including the landing in Brazil. Giovanni Francesco de Affaitadi, head of the Affaitadi merchant house in Lisbon, provided a similarly detailed account of Cabral's voyage based on information he obtained from a mariner.[17]

Maritime communities were also the places where chartmakers produced the most up-to-date charts. Angelo Trevisan di Bernardino, a secretary to the ambassador sent by the Venetian Signoria to Spain, wrote from Granada in August 1501 to Domenico Malipiero, a prominent Venetian, that he was seeking maps. Trevisan di Bernardino, more commonly known as Trevisan, claims to have struck up a friendship with Columbus, who was then in Granada and had in his possession a world chart. Columbus was in dire financial straits, Trevisan reports in his letter, and Columbus promised to have a copy of his chart made in Palos, a place, Trevisan explains, where sailors and men who sailed with Columbus lived. The chart, Trevisan writes, would be "beautifully made, and detailed, and in particular about the lands discovered." Columbus also promised Trevisan copies of his letters to the sovereigns Ferdinand and Isabela. Trevisan found it more difficult, however, to obtain the map showing the Portuguese route to India. He could obtain accounts of the voyage but not a map. He writes that "it is impossible to procure the map of that voyage [to Calicut] because the [Portuguese] king has placed a death threat on anyone who gives it out." Trevisan wrote again to Malipiero, probably in September 1501, about his success in acquiring a report of the Portuguese voyage to India, a report compiled by the Venetian emissary in Lisbon who had already written about the return of Cabral's ships. Trevisan writes, "In regard to the desire of Your Magnificence to learn of the voyage to Calicut . . . from day to day I am expecting Messer Cretico, who writes me that he has composed a small work. As soon as he arrives, I will see that Your Magnificence has part of it." The Venetian emissary—Messer Cretico—traveled to Spain from Lisbon in September 1501. During the next few months either he or Trevisan must have obtained maps, for Trevisan wrote Malipiero in December 1501: "If we return to Venice alive, Your Magnificence will see maps both as far as Calicut and beyond. . . . I promise you that everything has come in order." Although there is no record of the maps having arrived in Venice, the letter of Messer Cretico (also known as Il Cretico) was published in Venice in 1501.[18]

Vespucci returned from his second voyage to the Americas in 1502. In Lisbon, he began a letter to his patron Lorenzo di Pierfrancesco de' Medici, most likely in July 1502. Vespucci takes up right where he had left off in the letter that he had sent from Cape Verde, recounting that he crossed the ocean on a southwest by south course and reached land after sixty-four days at sea. Then sailing southwest for nine months and twenty-seven days—always along the coastline and crossing the equator and the Tropic of Capricorn—he estimated the total distance traveled at eight hundred leagues. In this letter Vespucci does not promise his patron a map but rather a text, which he calls his *Voyage*. He could not immediately send this "small work," however, because the king of Portugal had taken it from him.[19]

A chart once attributed to Vespucci is the *Portolan Chart* at the Huntington Library, also known as the *Portolan Chart (King-Hamy)*, dating to approximately 1502 (plate 5). Measuring 585 × 942 mm (about 2 × 3 feet), the world chart was originally nailed to a rod. Although now very faded, the chart was once richly painted in blue, red, green, silver, and gold. The toponyms are entered in red and black ink in a tiny script, with regional names in block capitals. The chart includes a latitude scale, a distance scale, and a double line for the equator.[20]

The *Portolan Chart* (*King-Hamy*) portrays the Atlantic and Indian Oceans much as Vespucci describes them. The western and eastern coastlines of Africa are depicted as the Portuguese reported them, following the voyages of Dias and da Gama. But moving from the eastern coast of Africa to India, the world chart shifts into a Ptolemaic view of Asia. Following Ptolemy's twelfth map of Asia, the island of Taprobane is placed centrally in the Indian Ocean and is surrounded by smaller islands. The equator runs through its southern region in the same place as on Ptolemy's map. The Magnus Sinus appears to the East of India, corresponding to the modern Bay of Bengal. The peninsula, with the city of Cattigara marked, extends into the Indian Ocean. Places that Vespucci names as coming from his conversations with Gaspar, such as Malacha, are located. The Indian Ocean is further illustrated with two ships.

In contrast, the Americas appear but only as a thin line. This coastline extends from the equator to a point well south, at a latitude more or less juxtaposed to the Cape of Good Hope. In northern South America there is a gap in the coastline north of the equator, and the line continues on the southern edge of the Caribbean, along what is today the northern coastlines of Venezuela and Colombia. No toponyms are given for the South American coastline, unlike the coastlines of Europe and Africa, which are well named. The islands of Cuba and Hispaniola are both labeled. The latitude scale, which likely represents the line of demarcation from the Treaty of Tordesillas, clearly places on the Portuguese side of the line Newfoundland, Labrador, and the Terra S[an]cta Crucis (Land of the Holy Cross)—Brazil.

Around the time Vespucci returned from his second transatlantic expedition, on which he claimed to have traveled eight hundred leagues along the coastline of South America, Alberto Cantino arrived in Cadiz. Cantino wrote his patron, the Duke of Ferrara, Ercole d'Este, several letters about events he observed in Spain and Portugal. From a letter dated 19 July 1501, the duke was among the first Europeans to learn that five ships had returned to Lisbon from India laden with spices, pearls, and other valuables. In a letter dated 17 October 1501, Cantino describes the return to Lisbon of a boat from the North Atlantic, carrying news of sailing among icebergs and of landing on a thickly forested land with a great river. These lands (Newfoundland or Labrador) were inhabited by peoples who lived from fishing and hunting, and fifty men and women from there were brought to Lisbon. Cantino claims to have "touched, and contemplated" them. In this same letter, Cantino reports that on 5 October 1501 a Spanish caravel arrived in Cadiz, returning from the Spanish Indies, carrying sixty Indian slaves, dyewood, and pearls. Cantino concludes his letter with a mention of a new policy of the Portuguese king: those accused of crimes could have their sentences commuted and receive a handsome payment if they went to settle these recently discovered places. Cantino thought, however, that few convicts sent to Santa Croce (Holy Cross), Brazil, would want to return because of its good airs and abundant and sweet fruits. There in fact, Cantino reports, five sailors had refused to reboard the king's ships and had stayed behind. In January 1502 Cantino wrote from Lisbon that the king of Portugal had newly invested Vasco da Gama as admiral of the Indies and that da Gama was about to leave Lisbon in command of an armada that would blockade the entrance to the Red Sea in order to attempt a Portuguese monopoly over the spice trade.[21]

Later that year, in 1502, Cantino obtained a world chart that now bears his name—*Carta del Cantino* (plate 6). Painted over six pieces of joined vellum, *Carta del Cantino* accentuates the Atlantic and its connections to India. The Atlantic Ocean is labeled as the "Oceanus Occide[n]talis" (Western Ocean) in the north and the "Mare Oceanus" (Ocean Sea) in the south. The Indian Ocean ("Oceanus yndias meridionalis") is connected to the western ocean (Atlantic) and to another ocean named the "Oceanus orientalis" (Eastern Ocean), which is clearly the still as yet unnamed Pacific. At the center of the world chart, placed over Africa, lies a magnificent compass rose from which radiate rhumb lines that cross the map. Hundreds of toponyms are lettered in, flags mark the realms of kings, and texts written in Portuguese describe lucrative trading opportunities available through maritime trade. Brilliant illuminations of cities, landscapes, animals, and compass roses further enhance the map.

Whether Cantino suborned an official inside the king's service to create a copy of the king's master map, or went to the best chartmaker he could find, or somehow obtained a map made for another, we do not know. For the world map that now bears his name, Cantino paid the sum of twelve gold ducats, or so he later claimed. Cantino left Lisbon with the world chart and arrived with it in Genoa in October 1502. This world chart had to have been finished in Lisbon only the month before, because it displayed knowledge that could not have been known in Lisbon earlier than September 1502. The South Atlantic islands of Ascension and St. Helena appear on *Carta del Cantino,* allowing the chart to be dated quite precisely, for João da Nova, who left Lisbon in 1501 and returned in September 1502, sailed along both of these islands, and it was he who brought information about them to Lisbon. In Genoa, Cantino delivered the world chart to Francesco Catanio, an established agent in the pay of the Ercole d'Este, the Duke of Ferrara. On the back of the world chart, Cantino wrote, "Navigational Chart to the islands newly found in India, given by Alberto Cantino to Signore Duke Ercole."[22]

Something about the world chart or about Cantino's commissioning of it (without the duke's permission) must not have pleased the duke, for he (or most probably his secretary) wrote to Cantino inquiring about its price. This letter does not survive, but Cantino's response from Rome, a letter dated 19 November 1502, does. In it, Cantino acknowledges receipt of the duke's letter and claims that he had paid the twelve gold ducats to obtain the map in Lisbon and that he was reimbursed by Catanio, the duke's agent in Genoa. Furthermore, Cantino implies that the map is worthy, stating "the map is of such a kind, and I hope, that [done] in this manner, it will please Your Majesty."[23]

A world chart made in Genoa, and known as the *Planisphère nautique,* so closely resembles *Carta del Cantino*, that it appears to be a direct copy (plate 7). The Genoese map bears the statement "opus Nicolay de Caverio Ianuensis" (work of Nicolay de Caverio, Genovese). Like *Carta del Cantino*, Caverio's *Planisphère* is also made from multiple pieces of vellum glued together: *Carta del Cantino* measures 105 × 220 cm (about 3 ½ × 7 feet) and is composed of six skins; Caverio's world chart measures 115 × 225 cm (about 3 ¾ × 7 ⅓ feet) and is composed of ten. They share a common view of the world, a common language (Portuguese), and many similarities in layout and textual legends. Both world charts have virtually identical legends for Brazil and Calicut. Further suggesting that Caverio copied Cantino's chart is that both display a similar error: the unthinkable omission of the Nile River. Both chartmakers employ the same iconographic imagery for the Americas—red birds and trees in Brazil, trees in Newfoundland—as well as several geographic and design choices for Africa, including a central compass rose, mountain ranges in North and South Africa, the castlelike trading post at São Jorge da Mina, and a lion in Sierra Leone. Still, the two world charts are not identical. Caverio added geographic information that is not found on *Carta del Cantino*, such as more place names along the coast of Brazil, and his choice of illuminations does not completely replicate those on *Carta del Cantino*. It would seem that, after Caverio copied Cantino's chart, he added more information from another source.[24]

As new views of the Atlantic became visible on charts, and as these charts moved from the maritime communities to cities within Europe, mapmakers saw them and were influenced by them. Some mapmakers began to reconfigure the place of the Atlantic Ocean on their world maps. For example, the Contarini-Rosselli map, engraved and printed in either Florence or Venice, dates to 1506. From a text box legend, Giovanni Matteo Contarini would appear to be the cosmographer, while Francesco Rosselli the artist who engraved the map. This map is believed to be the first printed map to show the New World. It might have been intended for an edition of Ptolemy's *Geography*, but the map appears to have been printed separately, and it exists today as a single copy. The map measures 630 × 420 mm (nearly 25 × 16 ½ inches), and it was printed on two sheets of paper. The creators of the map sought to integrate the new information about the

Atlantic, largely from Spanish voyages, into older Ptolemaic models—which emphasized Asia—and the still popular reports of such travelers as Marco Polo. The world map is drawn using Ptolemy's conical projection. The Atlantic emerges as vast and is connected to the Mare Indicum (Indian Ocean) in both the east and the west. The Spanish Indies appear prominently, as does the coastline of

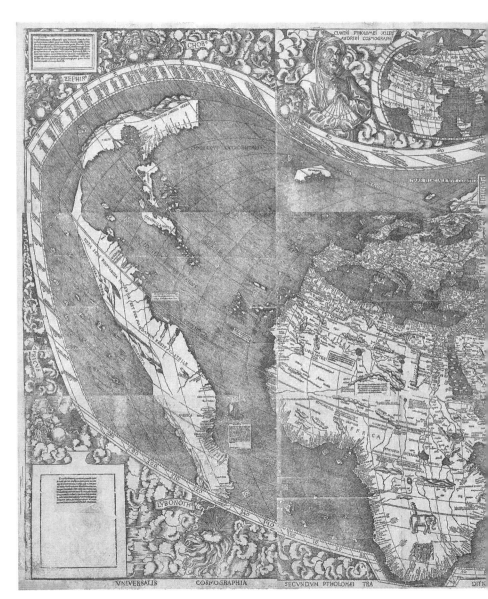

Fig. 2.2. Martin Waldseemüller, *Universalis Cosmographia*, 1507. Geography and Map Division, Library of Congress, Washington, DC.

northern South America. The eastern coastline of South America is given the name "Terra Crucis" (Land of the Cross), which refers to Brazil.[25]

In 1502 Martin Waldseemüller, Matthias Ringmann, and Walther (or Gualtier) Lud, all humanists in St. Dié under the patronage of Duke René of Lorraine, worked to produce simultaneously a large wall map, a globe, and a book.

The large wall map, Martin Waldseemüller's *Universalis Cosmographia*, was a massive undertaking. It was printed on twelve sheets of paper, making the entire map 128 × 233 cm (approximately 4 × 7½ feet). Waldseemüller modified one of Ptolemy's projections, and the result is the world flattened out in a kidney-like shape. Lines of latitude and longitude flow through the frame, and each continent is located under this graticule. Africa and Asia are large, and Europe stretches tight and taut across the northern shores of the Mediterranean. A large ocean separates Europe and Africa from the new islands and mainlands in the west. The Indian Ocean dominates the central south, while a third ocean—later to be named the Pacific—extends to the east beyond Asia at the right of the map and reappears in the west at the left of the map (figure 2.2).[26]

Unlike the *Universalis Cosmographia*, Waldseemüller's globe was small. None of the finished globes survive, but several copies of the globe gores do (figure 2.3). The gores are the printed sheets that would have been used to create the map on the globe. These surviving sheets consist of a printed world map designed in such a way that the paper could be cut and pasted over the ball. The world map is printed within twelve conjoined ellipses. The equator and the Tropics of Cancer and Capricorn appear, as do lines of latitude and longitude. Above and below the tropics the twelve ellipses sharpen into points, which, when joined, will create the North Pole and the South Pole.[27]

Waldseemüller's map and globe took advantage of the new advances in printing, which made it possible to create multiple copies. He later claimed the print

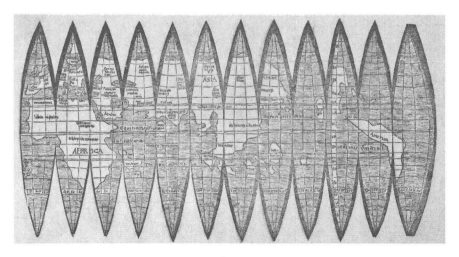

Fig. 2.3. Waldseemüller's Globe, 1507. Martin Waldseemüller, *The 1507 globular map of the world,* courtesy of the James Ford Bell Library, University of Minnesota, Minneapolis (*above*), and recreated globe (*opposite*).

run for a previous map, presumably that of 1507, to have been more one thousand copies. The map and globe presupposed a high level of literacy, but it was not necessary to be an educated humanist to purchase or to enjoy them. Waldseemüller had created objects that could be collected, even used as decoration. His was not a utilitarian chart for sailors and merchants, nor was it an elegant map printed inside an expensive book. His world map could be hung on a wall, and the globe set on a table.[28]

Artists began to produce maps that conveyed this new, expansive view of the Atlantic Ocean. During the sixteenth century, writes David Buisseret, "many painters could turn their hands to mapmaking, and many cartographers were also painters." In Florence and Venice, the maps available in Francesco Rosselli's shop were not exorbitantly expensive; even an unskilled workman could afford one. Not just artists made maps. Mapping became a "casually acquired skill" in the sixteenth century, argues art historian Svetlana Alpers, and "mapping, taken in its broad sense, was a common pastime." Artists and printers who were not necessarily trained as chartmakers nor as cosmographers began to make maps to sell to the public. Artists and printers who became interested in making maps borrowed from chartmaking and mapping traditions, and they developed their own approaches.[29]

Renaissance artists had created city and landscape views that held much in common with maps. Perspective views of Jerusalem, Rome, Venice, Florence, and other cities "are found everywhere in medieval and Early Renaissance art," writes art historian Juergen Schulz. By 1500, city views, such as that of Venice by Jacopo de'

Fig. 2.3. (continued)

Barbari, obscured the boundaries between maps and art. The sheer detail, size, scaling, and cartographic elements—such as windheads—of de' Barbari's *View of Venice* argue that it is both a map and a view of the city. Similarly, Leonardo da Vinci's plan of Imola (1502) is situated with a circle that mimics a magnetic compass with its "winds" or directions. As Genevieve Carlton argues, "artisan mapmakers" who were skilled artists "altered the way Europeans saw the world, and their success was in large part due to borrowing from the techniques of Renaissance artists and applying them to geographical works."[30]

Francesco Rosselli, the artist behind a world map drawn in a Ptolemaic projection by Giovanni Matteo Contarini, soon made his own world maps. Rosselli engraved a small world map in 1508 that used an innovative oval projection. The oceans dominate Rosselli's map, and because three surviving prints of this map were colored, it seems that the purchasers might pay more to have it finished with paints. The blues and greens of the colored versions, Genevieve Carlton emphasizes, argue that the world is "a sphere surrounded by clouds, with more water than land." The Atlantic, although not yet so named on Rosselli's map, is hardly marginal in the world.[31]

Cosmographers also sought to insert new information on the Atlantic Ocean into the corpus of maps prepared for editions of Ptolemy's *Geography*. The traditional set of maps drawn for an edition of the *Geography* included regional maps of Europe, Africa, and Asia, as well as a world map. While in the early editions the regional maps remained the same, the world maps began to evolve in order to include new information. For example, the 1508 edition of Ptolemy's *Geography*, printed in Rome, included a new world map created by Johannes Ruysch. Ruysch hailed from a noble family in Utrecht, studied at the University of Cologne, was ordained as a priest, and entered the Benedictine Order. In the early sixteenth century he was in Rome, where he worked as an artist. In 1507 the first edition of Marco Beneventano's edition of Ptolemy's *Geography* appeared with the twenty-seven Ptolemaic maps and the world map. The very next year, in 1508, a second edition was released with several new maps, including a new world map by Ruysch that was printed on two pages. Ruysch used a different projection—a "fan-shaped equidistant conic projection." In this projection, the Atlantic lies at the center of the world. Thus, in a major geographic publication for the time, in this edition of Ptolemy's *Geography*, the Atlantic is no longer marginal but is depicted as central to the world (figure 2.4).[32]

In less than a decade, creative and rapid changes in the spread of information, conceptualization of geography, and map design brought the Atlantic from the periphery to the center on world maps. We have no way of knowing how representative the surviving maps are, but we do know that they were made, pur-

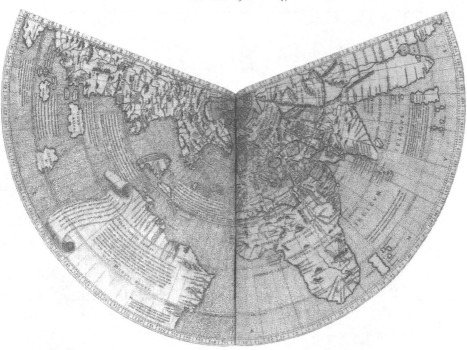

Fig. 2.4. Johannes Ruysch, *Universalior cogniti orbis tabula ex recentibus confecta observationibus*, 1508. In Ptolemy, *Geographia*. Rome: 1508. Courtesy of the John Carter Brown Library, Brown University, Providence, RI.

chased, and copied. As these maps became more familiar, they conveyed more than simply the extent and place of the Atlantic Ocean. Contained in them were the first arguments about the possibilities that lay in an interconnected Atlantic. Charts and maps did not simply portray the Atlantic as an ocean; rather they presented the opportunities in the Atlantic that could be explored, exploited, and claimed. These arguments for an Atlantic "world" first appear on charts, as we shall see in the next chapter.

CHAPTER THREE

Chartmakers

Juan de la Cosa's *Carta Universal* and *Carta del Cantino* (plates 4 and 6) are the oldest surviving world maps that present the Atlantic Ocean as a body of water bordered on four sides by continents. At the time, only a few would have seen these charts. Both were illuminated manuscript charts, either commissioned by or given as gifts to powerful lords. Their representations of the world, and especially of the place of the Atlantic in the world, were not at all familiar. In addition, each chartmaker had not simply charted more of the Atlantic Ocean but had represented it as a place for commerce, even evangelism and colonization. Mariners who lived in Cadiz or Lisbon and merchants who had invested in the maritime trade surely would have been quick to grasp the significance of this new presentation of the Atlantic. For although La Cosa and the chartmaker of *Carta del Cantino* created unique world charts, they did so using techniques that had mapped the Mediterranean since the thirteenth century. Common, well-known chartmaking techniques underlay *Carta Universal* and *Carta del Cantino,* even as each chartmaker introduced a bold new Atlantic World.

La Cosa crossed the Atlantic nine times. On his second voyage, he declared himself to be a "maestro de hacer cartas" (master chartmaker); he made *Carta Universal* after returning from his third voyage to the Americas in 1500. On *Carta Universal,* in the center of the Atlantic, labeled "Mare Oceanum" (Ocean Sea), is an icon of the Virgin Mary holding the Christ Child with two angels behind them. Circling the icon are a series of rings, triangles, and rays. Sixteen isosceles triangles are evenly spaced around the icon, and from their apexes extend thin straight lines that cross the ocean. On another ring, smaller triangles appear, and from them straight lines similarly range across the map. In all, thirty-two such rays leave the circle. The icon has been placed at the center of a compass rose (figure 3.1).[1]

This elaborate compass rose with a religious icon at its center is one indication that *Carta Universal* does more than simply present knowledge about geography. The compass rose, with its icon, invites the viewer to see the ocean in a certain light. Placed in the center of the Atlantic Ocean, the compass rose summons ships to sea by implying that navigation is possible and by reinforcing the belief that Mary offers protection to mariners. The triangles correspond to points of the magnetic compass, and the rays allow sailors to set a course (or at least to

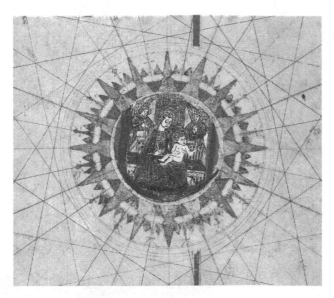

Fig. 3.1. Atlantic Compass Rose. Detail, Juan de la Cosa, *Carta Universal*, 1500, Museo Naval de Madrid & Biblioteca Virtual del Ministerio de Defensa, Madrid.

imagine setting one). More decorative compass roses appear on the world chart. Two in the Atlantic are smaller, each has a thin, prominent triangle that points north. Although they do not have a religious icon in their centers, they too convey the argument that the ocean is navigable. The rays from the sub-rose in the South Atlantic connect Africa with the newly discovered mainlands in the western Atlantic; while the rays from the sub-rose in the North Atlantic connect Europe with Africa and the not-yet-named Americas.

The compass roses are not simply decorative; they are part of the underlying layout of the world chart. La Cosa depicts the world in two hemispheres by using two invisible circles—one for the West and one for the East. The centerpoint for the western hemisphere in the chart lies under the icon of Mary, and although the circumference is invisible, it is marked with the smaller compass roses placed along it. A second, less prominent invisible circle, also with compass roses along its circumference, appears in Asia.

Two years later, the chartmaker of *Carta del Cantino* made the two-circle layout even more explicit through a more elaborate use of decorative compass roses. The chartmaker placed the most elaborate compass rose not in the center of the Atlantic but over Africa. And rather than placing a religious icon at its center, the chartmaker painted the smiling face of the sun. The message is different and not immediately clear. As La Cosa did, the chartmaker used two invisible circles to construct the world chart. Moreover, on *Carta del Cantino*, matching compass sub-roses stud the invisible circumferences like gems. In the chart's western hemisphere, the centerpoint of the invisible circle lies near the Cape Verde Islands. There is no compass rose drawn over it, most likely because if it had been drawn, it would have covered up the islands, which were important landmarks for Atlantic sailors. From the center point, thirty-two lines stretch directly across the Atlantic Ocean to the islands claimed by Spain in the Caribbean, to Europe in the northeast, to the west coast of Africa in the southeast, and to the coast of South America in the southwest. A second invisible circle, also with nearly identical compass sub-roses along its circumference, appears in the east. The centerpoint lies in India, and over it the chartmaker placed a sixteen-point compass rose. The rays from this rose stretch north into a huge and empty Asia, west to the Red Sea and the coast of East Africa, south to Antarctica, and east to China and beyond. By drawing these circumferences, as it is possible to do using a digital facsimile copy of *Carta del Cantino*, the placement of the decorative compass sub-roses on the perimeter of the invisible circles is evident (figure 3.2).

In the chart's western hemisphere, the decorative compass sub-roses circle the Atlantic, which the chartmaker labels as the "Oceanus Occide[n]talis" (Western Ocean) in the north and the "Mare Oceanus" (Ocean Sea) in the south. In the chart's eastern hemisphere, matching sub-roses also surround the Indian Ocean, which the chartmaker labeled "Oceanus yndias meridionalis" (South Indian Ocean). The invisible circumferences that enclose both hemispheres pass through the main compass rose—with the smiling face of the sun at its center—which

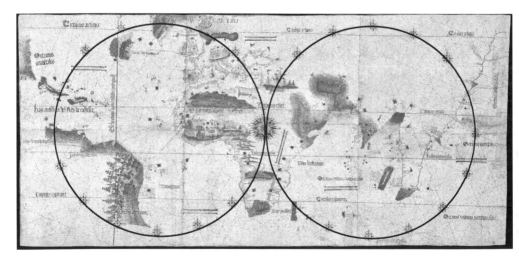

Fig. 3.2. Invisible Circles, *Carta del Cantino*, 1502. Digitally enhanced *Carta del Cantino*, Gallerie Estensi, Biblioteca Estense Universitaria, Modena, with permission from the Ministero per i Beni e le Attività Culturali e per il Turismo.

the chartmaker positioned in the center of Africa, between the equator and the Tropic of Cancer. This central compass rose not only stands at the center of the world but connects the two invisible circles that mark the eastern and western hemispheres (figure 3.3).

Through the invisible circles and their decorative compass roses, the chartmaker presents the world's oceans in a striking new way. Earlier map- and chartmakers— Martellus and La Cosa—had already shown the connection between the Indian and Atlantic Oceans, but on *Carta del Cantino* the Indian Ocean joins what the chartmaker labels the "Oceanus orientalis" (Eastern Ocean), which would later be named the Pacific. The compass sub-roses on the circumferences of the invisible circles suggest two maritime spaces: an Atlantic oceanic space and an Indian oceanic space in the west and east respectively. The oceans, particularly the Atlantic, are large, prominent, and interconnected. Yet the technique—placing a central

Fig. 3.3. Main Compass Rose, *Carta del Cantino*, 1502. Detail from *Carta del Cantino*, Gallerie Estensi, Biblioteca Estense Universitaria, Modena, with permission from the Ministero per i Beni e le Attività Culturali e per il Turismo.

point, drawing an invisible circle, extending networks of lines, and painting decorative compass roses—was not unique. Chartmakers had long used this format to create Mediterranean charts.

Although their origin is uncertain, and many questions remain about how they were made, charts had mapped the Mediterranean and Black Seas for a very long time. Tony Campbell identifies fewer than two hundred charts that have survived from before 1500, but these surviving examples nevertheless illustrate the essential features of the genre. Mediterranean charts often did include Atlantic coasts, especially as maritime commerce from Genoa and Venice extended to Atlantic ports, such as Lisbon, London, or Bruges. As noted in chapter 1, the *Atlas de cartes marines* includes a much larger depiction of the Atlantic than on the circular *mappaemundi*, and this information undoubtedly came from charts made by or known to the chartmaker.[2]

Whether called nautical, marine, sea, compass, portolan, or rhumbed, such charts presented the information valued, and often provided, by sailors, sea captains, pilots, and navigators. All charts are immediately identifiable by the network of lines—known as rhumbs—that extend from center points and radiate across the chart. The etymology of rhumb derives from Latin *rhombus,* but the word came into English from the Portuguese *rumo* or the Spanish *rumbo,* meaning line or direction. On navigational charts, the rhumb lines are prominent and inescapable. Mediterranean charts contained magnetic compass directions, scales, outlines of coastlines, and detailed place names. Charts taken to sea generally had little text written on them, with the exception of the place names, known as toponyms. Sailors, navigators, pilots, and captains who spoke different languages could understand a chart, and several people could study a chart at the same time. Charts provided a visual representation, but it is not entirely clear how they were used. Written logbooks, known as *portolanos* (Italian), *routeirs* (French), *roteiros* (Portuguese), and rutters (English), were also taken to sea. Such texts described ports, currents, shoals, the characteristics of the water, and distinctive landscape features—all essential information that sailors needed to reach and to return from their destinations.[3]

Historical charts fall into two very different groups, both of which were essential for the rapid emergence of the idea of an interconnected Atlantic World after 1500. The first types were utilitarian. These charts were nailed onto rods and rolled up when not in use; they could be taken to sea. Plain in their artistic design, they presented geographic information important to sailors—the outlines of coasts, reefs, shoals, and the names of prominent ports, landmarks, capes, bays, and river mouths. They also included a scale and magnetic compass directions. Charts had many limitations, and experienced captains relied on written logs, as well as past experience, to reach their destinations and return home.

Chartmakers produced a second, more elaborate, type of chart, for a different kind of patron—a merchant, lord, or prince. On these illuminated charts, chartmakers painted ornate and stylized cityscapes, peoples, kings, animals, and sometimes even (though rarely) monsters. Scenes painted on illuminated charts depicted events known or thought to have occurred in specific places. These high-end charts are similar to works of art in that they were commissioned by patrons, made to be aesthetically beautiful, and often envisioned as gifts.

Before 1500, the Atlantic began to appear on both plain and illuminated Mediterranean charts. The simpler navigational charts included the Atlantic Ocean as the western periphery extending beyond a carefully charted coastline in Western Europe or West Africa. The second type of chart—the illuminated chart—had more explicit arguments about the nature of the Atlantic as a new kind of space ripe with possibilities for commerce and exchanges. Still, both kinds of charts reflected the growing understanding of the importance of the Atlantic for trade and as a waterway that directly connected distant regions. As Matthew Edney has argued, "unless a region is first conceived of and named, it cannot become the specific subject of a map. Conversely, a mapped region gains prominence in the public eye."[4]

The *Portolan Chart of the Mediterranean*, chartmaker unknown, dated circa 1320–50, illustrates the basic elements of a late-medieval Mediterranean chart (figure 3.4). Measuring 43 × 50 cm, its most prominent feature is the visible network of lines—the rhumbs—that cross both the Mediterranean Sea and the land masses. Beneath this web of lines, the chart presents only the coastlines of the central and eastern Mediterranean, with part of the Black Sea. The chart is not enhanced with imagery, and it contains no textual legends. The only text on the chart is confined to place names—toponyms—written at right angles on the land side of the coastline. The chartmaker carefully lettered these toponyms in black ink but used red for the more important places, such as Rome. The chartmaker also used colors to differentiate the rhumb lines emanating from points located on the chart; these color-coded lines made it easier for mariners to read the compass directions. A scale is added near the bottom right corner.[5]

The earliest chartmakers were artisans whose family workshops were located in Mediterranean port cities, such as Genoa, Venice, and Majorca. Sailors, merchants, and captains purchased their charts from relatively few chartmakers. Historian Ramon Puchades i Bataller perceptively notes the existence of "many charts and many purchasers" but "few cartographers and even fewer ateliers [workshops]." Evidence from Venice suggests that Mediterranean chartmakers had extensive experience with the sea. Chartmakers were also shipowners, captains, and navigators, a "merchant-mariner class" in the words of historian Piero Falchetta. Other chartmakers were considerably poorer but may still have been

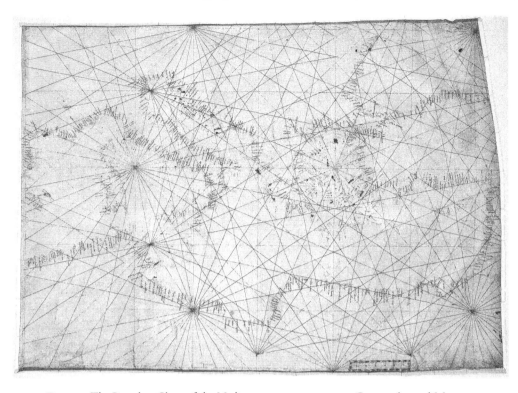

Fig. 3.4. The Portolan Chart of the Mediterranean, ca. 1320–50. Geography and Map Division, Library of Congress, Washington, DC.

to sea, or certainly lived among those who had. For example, Grazioso Benincasa, a prolific chartmaker from Ancona, spent much of his adult life at sea, serving as a ship's captain.[6]

A chart by Francesco Beccaria, also known as Franciscus Becharius, shows the Mediterranean, the North Atlantic Coast, the Black Sea, and the northwest African coast in 1403. Rhumb lines cross both land and sea, and toponyms, all written on the land side, emphasize the coastlines. Beccaria, known as "mestre de cartes de navegar" (master chartmaker) treats the Atlantic coastlines of Northwest Africa, Spain, Portugal, France, and Britain in the same manner as the coastlines of the Mediterranean and Black Seas. The chart is illuminated with several cityscapes—with Genoa the most prominent—as well as flags in several colors. The chart displays a clear integration between the waters of the Mediterranean and Atlantic, suggesting that seafarers easily accessed both. The known Atlantic archipelagoes appear, such as the Canary Islands off the coast of West Africa. Similarly, Grazioso Benincasa's chart of Europe, signed and dated 1470, shows a seamless incorporation of the Atlantic coastlines with those of the Mediterranean (figure 3.5). The coastlines with their toponyms are clearly the most important information recorded on the chart. There are few illuminations—a cityscape depicts Venice, and one full and two half compass roses appear on the chart.[7]

Chartmakers also made elaborate charts filled with illuminations that were not intended to be taken to sea. Letters about charts from the Datini archive, which preserves the correspondence and account books of a merchant from Tuscany, reveal the importance of such charts as gifts. Historian R. A. Skelton uncovered contracts for charts commissioned by an ivory and jewel merchant in Barcelona in 1399. The merchant intended to travel and sell his gems, and he hoped that by presenting illuminated charts as gifts to the kings of Aragon, Bordeaux, England, and Ireland they would guarantee him free movement in their kingdoms. In Barcelona, he commissioned four *mappaemundi* from two chartmakers. According to the archival papers, Jacome Riba of Mallorca was charged with the geography and Francesco Beccaria of Genoa with the illuminations. The latter claimed, as Skelton writes, to have painted "165 figures and animals, 25 ships and galleys, 100 fishes large and small, 340 flags on cities and castles, 140 trees."[8]

Letters recently found in the Datini archive describe other charts ordered from Majorcan chartmakers. From these letters it is clear that some charts

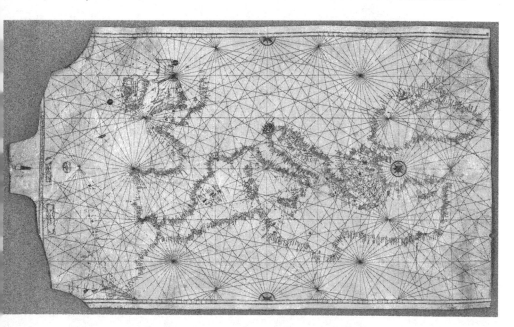

Fig. 3.5. Grazioso Benincasa, *A Portolan Chart of Europe*, 1470. The British Library, London.

commissioned or received were navigational charts, used by sailors, whereas others were illuminated charts intended as gifts. In one letter, the writer described a chart that arrived as "good, but it can only be used by sailors." This implies a navigational chart, such as that in figure 3.4. Several other letters refer to a chart ordered by Antonio delle Vecchie in 1395. Vecchie wanted the chart to be "beautiful and with banners," suggesting that he wanted an aesthetically pleasing final product. Another chart ordered by Luca di Biondo was commissioned as a gift. The chart he wanted to present to another merchant would likely be drawn from models in the chartmaker's shop but personalized to record the memory of a recent journey. These letters, argue historians Ingrid Houssaye Michienzi and Emmanuelle Vagnon, show that there was a commerce in charts in the late medieval period and that simple as well as elaborate charts were commissioned. Depending on their needs and desires, merchants purchased navigational charts for mariners, as well as illuminated charts to be contemplated, presented as gifts, and admired as souvenirs. Merchants displayed the illuminated charts in their homes, whereas mariners took the unadorned charts to sea.[9]

Mecia de Viladestes's *Carte Marine* (1413) is an example of a highly illuminated chart that nevertheless retains the essential elements of a chart. Made by a Mallorcan chartmaker, it, like the *Atlas de cartes marines*, is encyclopedic in the knowledge it presents; however, unlike the *Atlas de cartes marines*, it is not a *mappamundi* but focuses on the Mediterranean as it is situated within Europe, North Africa, and the eastern Middle East. The chart measures 850 × 1150 mm, and the outline of the Mediterranean coastline occupies the center of the chart, that of the Black Sea the upper right, and that of the eastern Atlantic the left. The Atlantic, redrawn and simplified in figure 3.6, illustrates the combination of rich illuminations with the basic underlying structure of central points, rhumb lines, and invisible circles characteristic of nautical charts. On this section of the chart, labeled as the "Mar Hocceano" (Ocean Sea), the Atlantic extends from West Africa to the Norwegian Sea. Center points where rhumbs intersect can be seen through the Atlantic, and the lines cross both land and sea. The toponyms extend down the West African coastline to Cap de Buyeter (Cape Bojador), while below Cape Bojador the West African coastline continues but is unlabeled. Offshore, two stylized islands mirror each other, and they are named Gades. The illuminations for West Africa recall the essential imagery and message of the earlier *Atlas de cartes marines*: a boat rigged with a lateen (triangular) sail, a man mounted on a camel, and the king of Meli (Mali) holding a golden nugget. In the northern Atlantic, England, Scotland, and Ireland are drawn, and an outline of Scandinavia is suggested. In the far north Atlantic, Viladestes adds a scene of a whale hunt in which sailors armed with a harpoon approach the whale from a

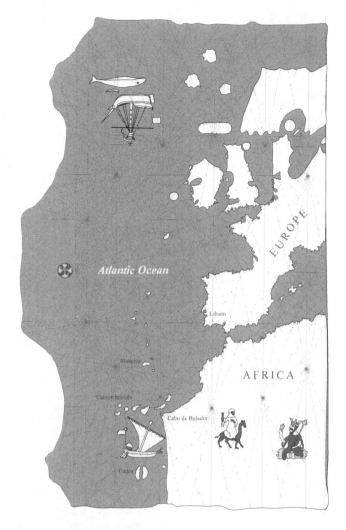

Fig. 3.6. An Illustrated Chart. Redrawn from Mecia de Viladestes, *Carte marine de l'océan Atlantique Nord-Est, de la mer Méditerranée, de la mer Noire, de la mer Rouge, d'une partie de la mer Caspienne, du golfe Persique et de la mer Baltique,* 1413.

small boat that has left a larger sailing vessel. The richly painted illuminations made it a luxury item. Yet the web of rhumb lines, which emanate from a central point in northwest Spain (with half of its subpoints in the Atlantic) point to the nautical nature of a chart.[10]

The Mediterranean had been a well-known body of water for centuries, and the early mapping of Atlantic coasts to the north and south of the Straits of Gibraltar could be incorporated using traditional chartmaking techniques. However, when chartmakers began to create charts that mapped mainly the Atlantic, they had to adapt some features of their craft because the Atlantic Ocean did not have the same kind of coherence as the Mediterranean Sea and navigating the Atlantic was a more complex proposition. Not only were distances greater, but currents and winds posed such challenges that even new designs for boats were needed.[11]

Chartmakers in Lisbon made some of the earliest Atlantic charts. The art and science of chartmaking arrived in Lisbon possibly in the fourteenth century but certainly by the fifteenth. Historians of historical cartography agree that the era of Dom Henrique (1394–1460), known in English as Prince Henry the Navigator, was critical in the development of chartmaking in Lisbon. Duarte Pacheco Pereira, who wrote an important work about navigation in the first years of the sixteenth century, states that Prince Henry brought a master chartmaker, Mestre Jacome, from Mallorca to teach his trade to chartmakers in Lisbon. The royal chronicler João de Barros, makes a similar statement in the first volume of his *Asia,* first published in 1552. Not many chartmaking shops would have opened in Lisbon, and these shops provided a modest living for very few families.[12]

Chartmakers in Lisbon served mariners and merchants who invested in an Atlantic trade with West Africa and the Atlantic island archipelagoes. Charts made in Lisbon visualized the coastlines of West Africa, located the Atlantic islands, and recorded information useful to captains and merchants. These earliest charts have all disappeared, and they are known only from a few scattered references. Gomes Eanes de Zurara, a fifteenth-century historian, made several comments on charts in his *Crónica de Guiné,* which describes the Portuguese exploration of West Africa. He describes a purported conversation between Dom Henrique and Gil Eanes, a navigator and captain who sailed past Cape Bojador in 1434. According to Zurara, Prince Henry spoke of the *agulha* (magnetic compass) and the *carta de marear* (sailing chart). Given the context of the remark, it indicates that while some captains under the patronage of Prince Henry sailed with charts and compasses, not all did. Because the Canary Islands had been known since classical times, and there had been many voyages to the archipelago, charts may not have been necessary. Sailing past Cape Bojador was another proposition as it lay past the known waters off the coast of West Africa. Zurara further comments on charts in additional passages, confirming that they were being made and taken to sea, and that they incorporated eyewitness information. In one passage, Zurara provocatively describes the chart as the means "through which all seas that can be navigated are ruled."[13]

More evidence for the charts made in Lisbon in the fifteenth century comes from legends on the world map discussed in chapter 1—*Il Mappamondo di Fra Mauro.* Knowledge obtained from charts directly influenced Fra Mauro's presentation of Africa; in one legend he writes that he had information from sailors "who were sent by his Majesty, King of Portugal, in caravels so that they might explore and see with their own eyes." Mauro further notes that these men had sailed two thousand miles beyond the Straits of Gibraltar and that they "have drawn new marine charts and have given names to rivers, gulfs, capes and ports, of which I have had a copy." Mauro's reputation as a mapmaker was such that the Portuguese crown commissioned a map from him, and sources indicate that the map was paid for and delivered in 1459.[14]

The oldest extant Portuguese chart is the *Carta nautica,* date unknown but falling between 1471 and 1482. The chart measures 617 mm × 732 mm (about 24 × 29 inches) and shows the Atlantic coastline from the Bay of Biscay to the Gulf of Guinea (plate 8). The chart contains all of the basic elements of the chartmaking tradition: rhumbs, coastlines, a scale, toponyms, and a few illuminations, including compass roses, flags, and cityscapes. The chartmaker did not embellish the lands depicted, with the exception of three cityscapes—Lisbon, Paris, and Ceuta. The Atlantic island archipelagoes—Azores, Madeira, Canary, and Cape Verde—are carefully placed and labeled. The Mediterranean does not appear, though its entrance from the Atlantic is mapped. It is clearly an example of an Atlantic chart.[15]

Jorge de Aguiar's illuminated chart of the Mediterranean, signed and dated 1492, is an example of a traditional Mediterranean chart expanded to include the Atlantic (plate 9). Known as the *Portolan Chart of the Mediterranean Sea,* it measures 80 × 104 cm (about 30 × 41 inches) and includes all of the Mediterranean and the Black Sea, but the Atlantic occupies almost one-third of the chart. The illuminations are not as prominent, nor as rich as on Viladestes's illuminated chart, but their presence reveals that chartmakers in Lisbon also produced higher-end charts for patrons who commissioned them. Over the land, Aguiar painted cityscapes for Jerusalem, Paris, Avignon, Barcelona, Valencia, Granada, Venice, Genoa, and Lisbon. In Africa, he painted a cityscape in the Atlas Mountains, and he included a stylized sketch of the recently established Portuguese trading post of São Jorge da Mina. Instead of animals or kings, Aguiar inserted views of the landscape, such as the mountains of Granada and Sierra Leone or the palm trees at Cabo Tres Palmas. In a style unlike the plain aesthetics of the *Carta nautica,* Aguiar carefully lettered in labels and added short textual legends. In Africa, Aguiar states "here is the range of the Montes Claros," to indicate the Atlas Mountains, and he labels Cape Verde and Sierra Leone. The Red Sea carries an explanatory legend: "This sea is not naturally red, but the land below it is."[16]

A few sixteenth-century Portuguese and Spanish texts describe how the early Atlantic charts were made. Pedro Nunes, the royal cosmographer of Portugal, published a text on the chart in 1537. In this text, Nunes describes the chart as the "best instrument that can be found for navigation and the discovery of lands." Nunes was concerned primarily with how to make charts more accurate for navigators, and he made only a few comments about the techniques used in the chartmaker's shop. In contrast, the chartmaker and cosmographer Alonso de Chaves, later appointed *piloto mayor* of the Casa de Contratación in Seville, described in detail the basic steps used to make a chart around 1535. Chaves viewed the chart as second in importance only to the compass in its utility for navigation. "The sailing chart," as he put it, "is to us like a mirror in which it represents to us the image of the world." Two decades later, the Spanish cosmographer Martin Cortés published a text in which he also specifies how to make a chart. For Cortés, the genius of the sea chart lay in its ability to show both the location of places and the distances between them.[17]

Chartmakers were artisans who did not learn their craft from books. Rather, they learned in the family workshops, where the craft was handed down from master to apprentice. Consequently, there are no accounts of how these early chartmakers of the Atlantic made their charts. Nevertheless, the written accounts by Chaves and Cortés provide many insights into chartmaking in Spain in the early sixteenth century. Even if the steps were not followed in the order described by Chaves and Cortés, it is likely that they were familiar to chartmakers a generation earlier in Lisbon. The few surviving charts made in Lisbon do suggest they followed the basic elements of the process, as described by Chaves and Cortés. They also illustrate how chartmakers adapted Mediterranean chartmaking techniques to include the Atlantic.[18]

The fifteenth- and early sixteenth-century chartmakers of Lisbon used the medium of paints and ink on vellum just as had their predecessors in Mediterranean workshops of the thirteenth and fourteenth centuries. To create vellum from calf, sheep, or goat skins involved a long messy process that would not have been done in the chartmaker's shop. Instead, chartmakers purchased the processed skins directly from artisans or dealers who specialized in the trade. Once the finished vellum was in the shop, the chartmaker laid it out on the workbench and examined it carefully. The two sides of the skin differed: the outer side was rough and brown, but the inner, which the chartmaker used for the chart, was white and smooth.[19]

Tools used in the workshop were the compass divider, a straightedge or ruler, quill pens, a knife, and brushes of various sizes. The chartmaker used the compass divider and ruler to measure. Quill pens made from the feathers of birds (such as geese) and brushes made from animal hair set into wooden handles transferred inks and colors to the vellum. The knife could trim the edges and also scrape away errors. Inks and pigments could be mixed as needed in the workshop or purchased from specialized artisans. Black ink was made from gallnuts, which were nutlike growths on oak trees caused by insects. When ground up into powder and dissolved in water, vinegar, or wine, the liquid could be combined with ferrous sulphate and gum Arabic to make ink. Red could be made from the shavings of brazilwood or from cinnabar or vermilion (both derived from mercury). Verdigris, a pigment naturally forming on copper and brass, became the basis for green. Blues came from several sources, the most ordinary from azurite, a commonly found stone that was beaten and ground into a fine powder with a mortar and pestle. Ultramarine provided a brilliant blue, but it was very precious and expensive because the lapis lazuli stone was found only in Afghanistan. Ochre came from soils rich in iron oxide that made a yellowish-brown pigment.[20]

With the surface prepared, the tools ready, and the inks and pigments at hand, the work began. First, the chartmaker secured the vellum to the workbench. When contemplating this blank surface, the chartmaker was not unlike an artist before a canvas. The chartmaker had an extensive repertoire of geographic knowledge, techniques, design choices, and imagery from the Mediterranean chartmaking tradition. But the chartmaker making an Atlantic chart often had new geographic information to include. There was a great deal available, and the chartmaker had to decide what to include and what to leave out, as well as how to visualize new places.

Cortés gives the chartmaker the following first steps: draw two perpendicular lines, which represent the North-South axis and the East-West axis. Then, place the compass at the intersection of the two lines and draw a huge invisible circle that encompasses all of the area of the chart. The chartmaker should then divide each quadrant in half to create eight winds, then each eighth in half to create sixteen winds, and finally each sixteenth in half to create thirty-two "winds." These "winds" referred to the directions from which the winds blew when a sailor was at sea, but on the chart they became the rhumbs that corresponded to the directions of the magnetic compass. Cortés does not explain how the chartmaker is to measure the angles. For example, was a protractor used, or did the chartmaker manipulate just the compass and the straightedge to create the angles? In any case, the process required an understanding of basic geometry as well as a deft manipulation of the instruments. Chaves provides similar but less detailed instructions and states that once the circle is established and the web of lines set out, the chart is "rhumbed" (*arumbado*).[21]

The result of this process can be seen clearly on the *Carta nautica*. The chartmaker placed the centerpoint within West Africa, due south of Cabo de Gué, a Portuguese trading post. The circumference created from this central point is not visible on the chart, but where it passes can be worked out because it is where additional intersection points lie, and some of these sub-points are decorated with compass roses. The anonymous chartmaker did not draw all thirty-two rhumbs from the central point; instead he drew only sixteen. However, other sub-points on the chart have all thirty-two points and rhumbs. The invisible circle extends to the edges of the vellum and encloses the geographic information on the chart (plate 8).

On Jorge de Aguiar's *Portolan Chart of the Mediterranean Sea*, the center point is readily apparent—it is decorated with a large compass rose—and where the circumference of the invisible circle lies can be determined by looking for the compass sub-roses. Aguiar placed the central point for the chart in North Africa, near Algiers, and over it he painted a sixteen-point compass rose. Six sub-roses appear over intersection sub-points on the invisible circumference. The invisible circle encloses most of England and Ireland in the north, most of the western Mediterranean in the east, and just past Cape Verde in the south. In the west, it cuts through the Atlantic where two intersection sub-points and two decorative compass sub-roses are prominent (plate 9).

These center points and invisible circles are found on earlier Mediterranean charts, where they too enclosed the most important geographic information on the chart. For example, a chart by Gabriel de Vallseca, dated 1447, has its centerpoint in Greece. Chartmakers in Lisbon found their mapping tradition flexible as they sought to include the Atlantic. Moving the center point, for example, allowed them to include more of the Atlantic, as necessary. A Mediterranean chart made in 1500, the anonymous *Alte Welt* chart, has its center point in North Africa near Algiers, similar to the placement of the center point on Aguiar's *Portolan Chart of the Mediterranean Sea* (plate 9). For Atlantic charts of West Africa, chartmakers placed the center points farther south or farther west. The centerpoint for Pedro Reinel's *Portulan* (plate 10) lies southeast of Cabo de Gué, as does the center point for the *Carta nautica* (plate 8).[22]

Chartmakers made another adjustment when creating Atlantic charts, which had to do with the orientation of the vellum. When specialized artisans processed animal skins into vellum, they typically retained the skin corresponding to the animal's neck. When the vellum arrived in the chartmaker's shop, it had a "neck" or a tab on one side. Chartmakers creating a Mediterranean chart placed the "neck" so that it lay on the left, or in the West, where it tapered off into the unknown Atlantic. It also made sense to take advantage of the rectangular shape of the animal skin to show the east-west axis of the Mediterranean. On his *Portolan Chart of the Mediterranean Sea*, Aguiar positioned the vellum so that it was wider than it was tall (plate 9). This made sense because his chart showed the full Mediterranean. Chartmakers who created Atlantic charts might choose to turn the vellum such that it was taller rather than wider. The "neck" of the chart then appeared on the top, pointing north. Having the neck on top allowed more of the European and African Atlantic coastline to appear. This decision to rotate the neck visually reinforced a north-south axis for the Atlantic.

At some point in the life of the *Carta nautica*, it seems to have been trimmed, for there is no "neck" on the chart. However, because the size of the chart is longer rather than wider, the neck originally would have been on top, pointing north. The chartmaker Pedro Reinel clearly did turn the animal skin and position north at the top on *Portulan* (plate 10). By placing the neck of the vellum at the top, Reinel included less of the Mediterranean and omitted entirely the Black Sea. Turing the vellum gave Reinel more room to map Atlantic coastlines, and the north-south axis of the Atlantic is obvious.[23]

Two chartmakers inserted additional pieces of coastline onto their charts, in order to expand the geographic information that could be included. They placed these sections of coastline over the interior of Africa. On the *Portolan Chart of the Mediterranean Sea*, Aguiar extended the coastline from Sierra Leone to São Jorge da Mina by adding the outline of the coast, with toponyms, a landscape illustration, three palms, and a cityscape, over West Africa (plate 9). Historian Alfredo Pinheiro Marques calls this technique, which is also visible on Pedro Reinel's *Portulan* (plate 10), "an ingenious resource on the part of each cartographer," for it allowed him to "represent the scale and relative positions of the coasts and rhumblines to which they were used and go on representing the same pattern of lines and coasts, with the compass-roses in the same places, and still use the same general scale." Like Aguiar, Reinel placed a section of the West African coastline over the interior of Africa. This additional coastline extended an even greater distance— from São Jorge da Mina to the Zaire River.[24]

Chartmakers grappled with how to incorporate new regions and vast distances, especially in the South Atlantic. At first, as chartmakers mapped the coasts of Northern Europe, from Gibraltar to London, for example, following the techniques used in the Mediterranean worked well enough. Similarly, as chartmakers mapped the coast of West Africa, using the Mediterranean approaches would also be effective. However, problems emerged with the Mediterranean method as distances increased. Magnetic declination slightly affected the readings taken from a compass, and compasses themselves varied in quality. Combining information from various Atlantic voyages onto a single chart required standardizing compass readings, which was not always possible. Errors accumulated, making some information on the Atlantic charts unreliable and even contradictory.[25]

Chartmakers began to use latitude scales on Atlantic charts to improve their reliability, especially as they began to map places south of the equator. The astrolabe became an essential maritime instrument in the South Atlantic (where the North Star cannot be seen). The astrolabe measured the height of the sun at high noon, and these readings, when combined with tables of solar declination, resulted in good calculations of latitude. To incorporate this information onto a chart, a chartmaker added a scale of latitudes. Once a chartmaker entered a latitude scale, places with measured latitudes, such as those taken by mariners with the astrolabe, could be plotted. These techniques enabled chartmakers to include observations of latitudes at coastal ports at key landmarks—such as the *padrões* erected along the coast of South-Central Africa—onto their charts.[26]

According to Chaves, after placing the center point, the chartmaker entered the latitude scale (if one was to be used). Only after this basic layout would the chartmaker began to enter the geographic information. First came the coastlines. For the chartmaker, the coastline must be placed as accurately as possible, but design must take into account the need to record hazards, such as shoals, sandbars, and reefs. Essential too were key landmarks for sailors, such as islands, major capes, river mouths, bays, and harbors.

Chartmakers copied coastlines from chart to chart. Cortés refers to the *padrón*—a pattern chart—that the chartmaker could trace using thin, transparent paper. Puchades i Bataller suggests that silk might have been used in medieval times. Family workshops most likely had pattern charts on hand that they used over and over until they wore out or were judged out of date. To transfer information from the pattern chart to the freshly prepared vellum, a method known as "pouncing" could be used. When pouncing, the chartmaker pricked tiny holes in a pattern chart and then dusted the pounce (with lead powder, chalk, or carbon) over the top. By lightly tapping, the pounce passed through the holes and left a line on the underlying vellum. Once the pounce had transferred to the new chart, Cortés gave the last step—tracing the coastline with a fine pen and, once the ink had dried, "with a bit of bread" cleaning off the pounce, "leaving the coastline entered with ink onto the chart."[27]

Chartmakers could only copy a coastline from a pattern chart if making multiple copies of the same chart, or if the scale on the new chart was the same as on the old. Resized charts or charts that introduced new coastlines required a different technique. Chaves describes how this is to be done. The chartmaker must first locate on the original map a major geographic feature, such as a cape, and then carefully determine its latitude. Then this latitude must be transferred to the equivalent place on the new chart. The next step is to locate another prominent geographic feature, such as another cape, not too far distant. The chartmaker must also note its latitude and also determine its position relative to the first point. Chaves recommends that "one must have a very good pair of compass dividers"—the instrument chartmakers used to measure distances. Then, Chaves states the chartmaker works as an artist designing the coastline between the two known points. In Chaves's time, the chartmaker could use lead that could be "later inked in using a fine pen." Following Chaves's description, the exact edge of the coastline did not have to be geographically accurate; instead, the latitudes of the major features that the sailors would encounter, such as capes, must be located as precisely as possible. Between these points, the coastline could be approximated and stylized.[28]

Charting coastlines was key to how chartmakers mapped the Atlantic as an ocean. The coastline is an essential feature of ocean geography, and knowing where coastlines lay was indispensable for navigation. Conventions for how to chart a coastline seem to have been adopted and standardized between chartmakers' shops, such that the same sections of coastlines appeared on different charts in fundamentally similar ways. For example, the section of the West African coast—that lying between Cabo Branco and Cabo Verde—is virtually identical on five charts made in Lisbon: the anonymous *Carta nautica* (plate 8), Aguiar's *Portolan Chart of the Mediterranean Sea* (plate 9), Pedro Reinel's *Portulan* (plate 10), *Carta del Cantino* (plate 6), and Pedro Reinel's *Portulan (Atlantik)* (plate 11). The coastline geography is strikingly similar: shoals are marked south of Cabo Branco, which, as described by Alvise Ca' da Mosto, created a dangerous approach to the trading post on the island of Arguim. On all five charts, Cape Verde extends prominently into the Atlantic, and on all charts, the coastline is stylized. Differences are visible, however, in how the chartmaker chose to symbolize the coastline, islands, and landscape. Comparing two charts (figure 3.7) illustrates how chartmakers presented the same geographic information in slightly different ways. Aguiar and the anonymous chartmaker of *Carta nautica* placed a Portuguese flag at Arguim, right below Cabo Branco. Each marked the shoals with a series of dots and crosses, and Aguiar even labeled them. The coastline is stylized on both charts and highlighted in a similar way. Aguiar and the anonymous chartmaker drew a thin line then added a light pigment along the landside. The chartmakers also made choices in rendering other geographic features, such as islands, river mouths, bays, and promontories. Aguiar painted islands in both blue and red, richly illustrated Cape Verde with green, and selected a bright blue for the Senegal River (Rio Çanaga). The *Carta nautica* is plainer.

After drawing the coastline, Chaves gives as the chartmaker's next step the adding of names either on the landside or on the ocean side, depending on where there was space. Cortés writes that the chartmaker first enters the most important place names in red, followed by those of lesser importance in black. Scholars consider the toponyms—the names given to ports, bays, capes, river mouths, villages, towns, and cities—as some of the most important information on a chart. Historians can use toponyms to date charts, to trace the evolution of a particular place, and to determine which charts chartmakers copied. Toponyms

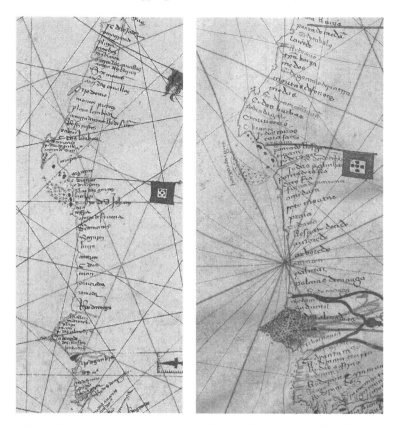

Fig. 3.7. The West African Coastline on Two Charts. Created in ArcGis from georeferenced, *Carta nautica*, ca. 1471–82 (*left*), and Aguiar, *Portolan Chart of the Mediterranean*, 1492 (*right*).

also contain in their naming patterns the history of voyages, for they may recall events that took place, or record a date by using names taken from saints' days from the Christian religious calendar. Toponyms convey features of the landscape deemed important or unique. Chartmakers had a good deal of choice in the selection of toponyms, and they could modify names, change spellings, transcribe names into other languages, and miscopy them.[29]

Toponyms inserted for the Guinea coast between Cabo Branco and Cabo Verde on the five charts made in Lisbon discussed earlier—the anonymous *Carta nautica* (plate 8), Aguilar's *Portolan Chart of the Mediterranean Sea* (plate 9), Pedro Reinel's *Portulan* (plate 10), *Carta del Cantino* (plate 6), and Pedro Reinel's *Portulan (Atlantik)* (plate 11)—are similar. All are written in Portuguese and in a clear hand. Cabo Branco, Arguim, Rio de São João, Rio da Çanaga, and Cabo Verde appear in red ink across all five charts; clearly, they are accepted as the more important places. Aguiar's *Portolan Chart of the Mediterranean Sea* has the most toponyms between Cabo Branco and Cape Verde at twenty-eight, while Reinel's *Portulan (Atlantik)* chart has the least, at twenty. Not all of the toponyms refer to specific places; some simply convey a sense of the land. For example, among the toponyms on all five charts are entries such as *médão/médãos* (sand dunes), *praia* (beach), *moutas* (thickets), and *montes* (hills). The slight variation in names for the same places points to the fact that although there was a general acceptance of naming patterns, the names were not completely standardized. Abbreviations, spellings, and slightly different forms appear for what are clearly the same places.

The toponyms tell an important part of the story of mapping the Atlantic. Toponyms assert the right to name and rename. The act of naming created analogies, documented events, projected bias, and imposed order, all of which influenced the interpreting of geographic features by the map viewer. How chartmakers kept track of the names is uncertain. Possibly the toponyms came from pattern charts that the shop had on hand, or the names were stored in copybooks, so that new names could be easily added as information arrived. As names were copied from one chart to the next, names were imposed on African regions, even as slight differences appeared in spelling or as some names were miscopied. Over time, chartmakers established standardized names through their repetition. The basic toponyms required for the coast between Cabo Branco and Cape Verde were clearly well established by 1500, suggesting this way of naming was repeated for coastlines throughout Africa.

As detailed as this stretch of coastline is between Cabo Branco and Cape Verde on the five charts discussed here, sailors would not rely solely on charts, even though each chart shows with crosses and dots where the dangerous waters lay. Duarte Pacheco Pereira's *Esmeraldo de Situ Orbis*, which preserves many of the characteristics of a logbook or *roteiro,* clarifies how a written text reinforced essential details that charts illustrated visually. In the description of the coastline between Cabo Branco and Cape Verde, Pereira writes that the shoals of Arguim are thirty leagues long and twenty wide, and very treacherous. Captains wishing to sail to one of the river mouths of Guinea, he wrote, should, once they are in sight of Cabo Branco, set a course south-southwest for ten leagues, next sail one hundred leagues due south, and then head south-southeast to reach the Angra das Almadias. The land between the São João and Senegal Rivers, Pereira writes, is completely covered with sand, and the stretch should be avoided because of the many sand bars and reefs.[30]

After the coastline and the toponyms had been laid down on a chart, Chaves states that only then does the chartmaker add characteristics of the lands, such as mountain ranges, cities, and kingdoms. Islands—oversized to make them visible—could be inserted late in the process, as well as markings for shoals, sand-bars, and reefs. Finishing touches were the final inking in of the rhumb lines, often color coded, and the major parallels (the equator, the Tropics of Cancer and Capricorn, and the Arctic Circle). Also, late in the preparation the chartmaker could add decorative compass roses over the central and sub-points, as well as the lettering in of any labels or legends.[31]

Ubiquitous on charts is the compass rose. Drawn over the central and sub-points, chartmakers drew full and half roses on their Mediterranean and Atlantic charts. Chartmakers also made the magnetic compasses carried on board ship. Chartmaker Pedro Reinel's title under King João II (1481–95) was *mestre de cartas e agulhas de marear* (master of charts and sailing compasses), as was his son Jorge Reinel's under King João III (1521–57). The chartmaker Lopo Homen had the same title under King Manuel (1495–1521) and King João III. Chaves and Cortés give detailed directions for making the magnetic compasses to be taken to sea. Because chartmakers also made these magnetic compasses, the elaborate compass roses visible on charts may well have matched the design and style on the dial of the physical compasses. Charts and compasses could be made as matching sets, thereby encouraging the use of the chart for navigation.[32]

According to Cortesão, Astengo, and other scholars of historical cartography, decorative compass roses on charts were not needed for navigation at sea, and most charts made to be taken to sea probably would not have had them. "Their appearance seems to be directly linked to the growing taste for cabinet nautical charts, which had either a didactic purpose or no nautical purpose at all," Astengo writes. If the most beautiful charts were the most likely to survive, and compass roses were part of their illumination, then the presence of compass roses is over-represented on surviving charts. Yet Chaves notes the custom of decorating charts with the central rose, as well as the creation of roses over the sub-points.[33]

On the Atlantic charts, the compass roses play a more prominent role in the aesthetics of the chart because of how they frame the geography of the Atlantic. Compass roses had always been part of the chartmaker's design kit for making Mediterranean charts, but on Atlantic charts, given the amount of space devoted to the ocean, the compass roses became more visible. On the *Carta nautica,* the chartmaker created two styles of roses: one for the central compass rose and another for the smaller sub-roses that appear on the invisible circle. The central rose is larger than the sub-roses and has sixteen points. The sub-roses are virtually identical, each with eight points, colored in red and black, and topped with a similar black helmet to mark North. The sub-roses form a matching set. The colors in the central rose repeat those of the sub-roses, and the top piece on the central rose matches those of the sub-roses. (figure 3.8).

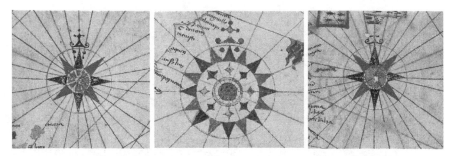

Fig. 3.8. Compass Roses on *Carta nautica,* 1471–82. Detail from *Carta nautica,* Gallerie Estensi, Biblioteca Estense Universitaria, Modena, with permission from the Ministero per i Beni e le Attività Culturali e per il Turismo.

On his *Portulan (Atlantik chart)* circa 1504 (plate 11), Pedro Reinel placed the centerpoint in the middle of the Atlantic, and he drew a sixteen-point rose over it. Thirty-two rhumb lines extend from the central point out to sub-points on the invisible circle. These sub-points, six of which are decorated with compass roses, three with half roses, encompass the Atlantic. Reinel's use of a visually prominent central compass rose, as well as conspicuous sub-roses around the invisible circle, defines the Atlantic as a central geographic feature of the chart.

Beyond the compass roses, Atlantic charts created in Lisbon carried little in the way of visual imagery. On the *Carta nautica* (plate 8), the compass roses and simple cityscapes are the only illustrations. Chartmakers paid more attention to illustrations when preparing higher-end illuminated charts, such as Aguiar's (plate 9) or Reinel's (plate 10). On illuminated charts, patrons and illuminators might have requested or suggested designs to the chartmaker. Aguiar included more cityscapes and painted them with individual touches suggestive of their location: Lisbon appears as a city on a hill along a river, Venice floats in a watery space, and Genoa curves around its distinctive bay. Aguiar also emphasized the differences between landscapes. He highlighted Cape Verde in green; he painted the landscape of Sierra Leone as brown hills with trees; and he placed three tall palms to mark Cabo das Palmas (figure 3.9).

Reinel's illuminations on his portolan of Africa (plate 10) did not include cityscapes or differentiated landscapes; instead, he emphasized flags and placed a carefully rendered lion in Sierra Leone. The way Reinel painted the lion is revealing, for the lion is not represented as if it were part of the natural fauna; rather it stands, as a human would, pointing into the interior with one paw and holding a Portuguese flag in the other. The lion with its flag symbolizes the Portuguese interests in Africa.[34]

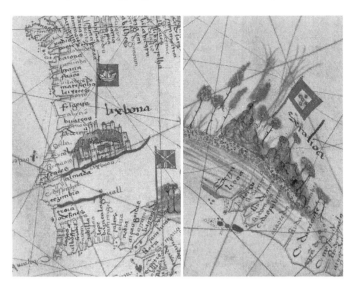

Fig. 3.9. A Cityscape and a Landscape on Aguiar's *Portolan Chart*, 1492. Details from Aguiar, *Portolan Chart of the Mediterranean Sea*, Beinecke Rare Book and Manuscript Library, Yale University, New Haven, CT.

Thus, the Atlantic emerged gradually on charts, beginning with those made primarily for Mediterranean sailing. Moving the center point and its invisible circle, extending rhumb lines to distant shores, using the same design for coastlines, recording scores of carefully lettered toponyms, and the placement of decorated compass roses all allowed chartmakers to include new regions and to present them as accessible. In this way, the Atlantic Ocean began to appear not only as a geographic entity—a body of water—but also as a navigable place that invited trade, exploration, colonization, and religious missions. In 1500 and 1502, *Carta Universal* and *Carta del Cantino* used the traditional techniques of chartmakers to make the argument that an Atlantic world has opened in the West. This idea would soon be visible far beyond the maritime communities when it appeared on printed maps.

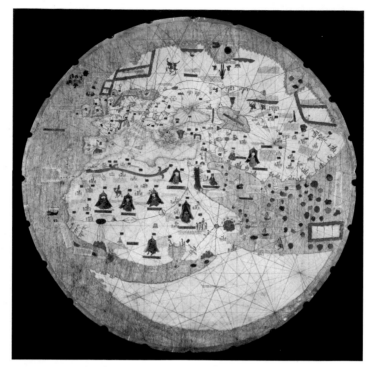

Plate 1. *Mappamondo Catalano estense*, 1450–60. Gallerie Estensi, Biblioteca Estense Universitaria, Modena, with permission from the Ministero per i Beni e le Attività Culturali e per il Turismo.

Plate 2. *Il Mappamondo di Fra Mauro*, 1450. Biblioteca Nazionale Marciana, Venice.

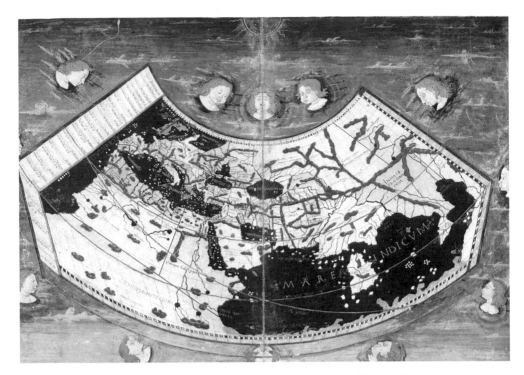

Plate 3. Nicolò Germano, *Planisfero*, 1466. Gallerie Estensi, Biblioteca Estense Universitaria, Modena, with permission from the Ministero per i Beni e le Attività Culturali e per il Turismo.

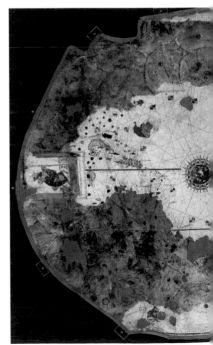

Plate 4. Juan de la Cosa, *Carta Universal*, 1500. Museo Naval de Madrid and Biblioteca Virtual del Ministerio de Defensa, Madrid.

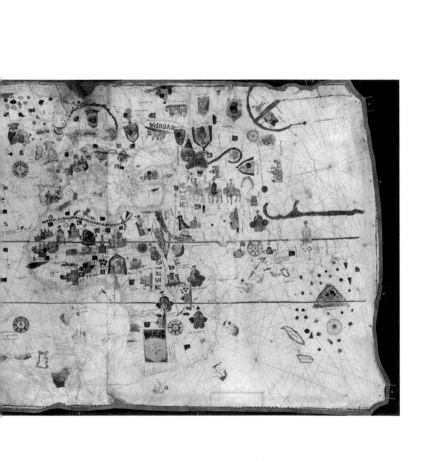

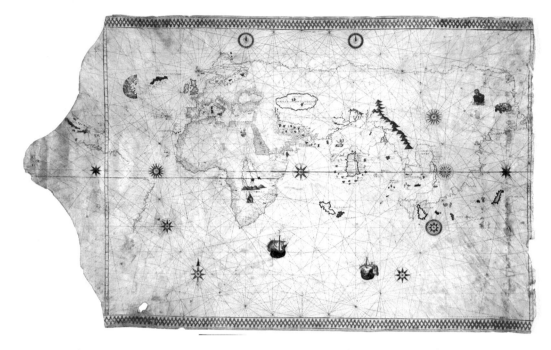

Plate 5. Portolan Chart (King-Hamy), ca. 1502. HM 45, The Huntington Library, San Marino, CA.

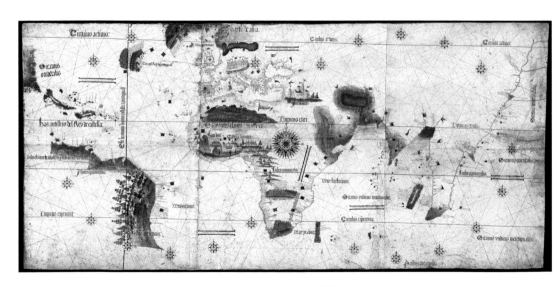

Plate 6. Carta del Cantino, 1502. Gallerie Estensi, Biblioteca Estense Universitaria, Modena, with permission from the Ministero per i Beni e le Attività Culturali e per il Turismo.

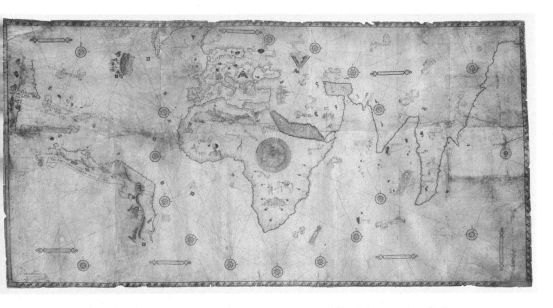

Plate 7. Nicolay de Caverio, *Planisphère nautique*, 1506. Bibliothèque nationale de France, Paris.

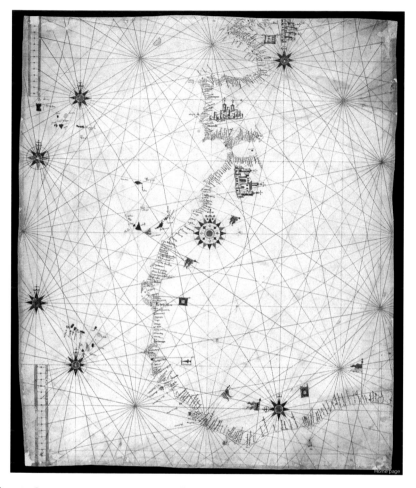

Plate 8. Carta nautica, ca. 1471–82. Gallerie Estensi, Biblioteca Estense Universitaria, Modena, with permission from the Ministero per i Beni e le Attività Culturali e per il Turismo.

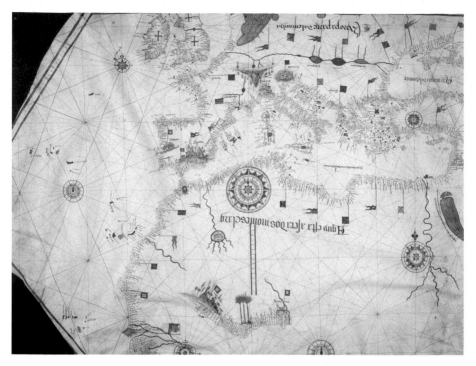

Plate 9. Jorge de Aguiar, *Portolan Chart of the Mediterranean Sea*, 1492. Beinecke Rare Book and Manuscript Library, Yale University, New Haven, CT.

Plate 10. Pedro Reinel, *Portulan*, ca. 1485. Archives départementales de la Gironde, 2 Fi 1582 bis, Bordeaux.

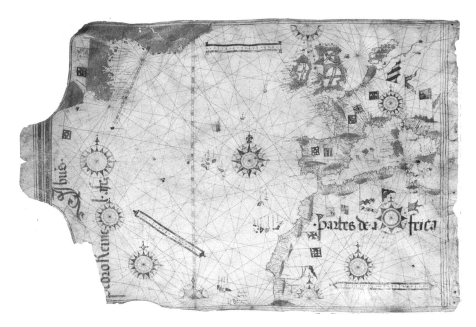

Plate 11. Pedro Reinel, *Portulan (Atlantik),* ca. 1504. Bayerische Staatsbibliothek, Munich, Cod.icon. 132.

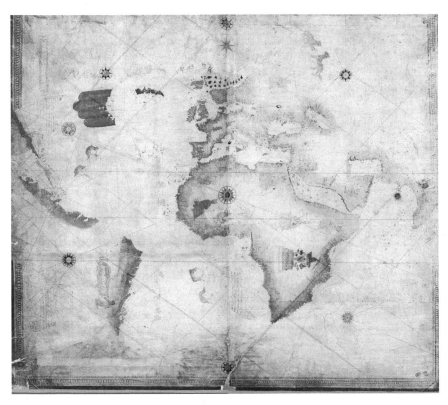

Plate 12. Portolan (Weltkarte), 1502–6. Bayerische Staatsbibliothek, Munich, Cod.icon. 133.

CHAPTER FOUR

From Manuscript to Print

Manuscript charts had a small circle of viewers. Chartmakers crafted each chart—whether plain or illuminated—for a specific voyage or patron. The same was true for manuscript maps made by Renaissance artists and cosmographers. Even though, as art historian David Buisseret has written, "many painters could turn their hands to mapmaking, and many cartographers were also painters,"[1] hand-painted maps still circulated among very few. The mapmaker who created a printed map, on the other hand, had a very different public. The potential audience was large, distributed over a wider area, and broad in interests. Printed maps became influential after 1500, and for many living in Western Europe who had never been to sea, nor would ever go, their visual comprehension of the Atlantic Ocean came from printed maps. Not only did the early printed maps spread the ideas about the world of the Atlantic known only in maritime communities far and wide but printed maps introduced new knowledge.

The years immediately after 1500 mark a new visibility for the geographic space of the Atlantic Ocean, which was part of a longer and more significant transformation in the use, availability, and importance of maps. Gradually, over the course of the sixteenth century, maps became familiar to many in Western Europe. "In the England of 1500," P. D. A. Harvey writes, "maps were little understood or used. By 1600 they were familiar objects of everyday life." Robert Karrow describes a "sea change in European consciousness of the possibilities of the map" that became visible by 1600. Before 1500, Denis Wood argues that, give or take a few exceptions, *people didn't make maps,* but by 1600, maps had become indispensable, not only in Europe but around the world. Printers propelled this change. Whereas maps had been a marginal genre in medieval Europe, Karrow estimates that by 1600 more than 1.3 million printed maps were in circulation. In the heartland of Western Europe where the printing revolution took place, he reckons that there would have been one map for every 720 persons in 1500; by 1600, there was one for every 4. The map trade brought about a "radical change in the patterns of ownership of maps," writes David Woodward, for the "modest prices of prints compared to original artworks enabled the middle classes to enjoy a consumerism in collecting that had been previously reserved for the nobility."

The explosion of printed maps, which were illustrated with representations of peoples and places, writes Surekha Davies, invited the "comparative contemplation of human civilizations." Moreover, by the end of the sixteenth century, Davies argues, "illustrated cartography was one of the scholarly achievements that emblematized European intellectual achievement and artisanship."[2]

Printed maps predate 1500 in Western Europe. Like the manuscript *mappae-mundi* made using the medieval T-O layout or those made with projections recommended by Ptolemy, the early printed maps of the world gave little importance to the Atlantic. Yet print opened up a new medium for maps, and as printers became familiar with maps—whether to be bound in books or sold as separate sheets—printed maps became increasingly common. The first printed edition of Ptolemy's *Geography* to contain maps was released in five hundred copies at Bologna in 1477. The maps in this edition, as well as in another that came out in Rome the next year, were engraved and printed from copper plates. Yet another edition appeared in Rome in 1480, and two years later, one in Florence. Using a different process—woodblock printing—Leinhart Holle printed the first edition of Ptolemy's *Geography* outside of Italy, in Ulm, Germany, in 1482. An edition of Ptolemy's *Geography* was costly and challenging to produce. Printers faced long lists of place names, each with their specific coordinates of latitude and longitude. The maps required particular attention, for an artist or a cosmographer would have to prepare them. Copperplate engraving allowed the artist to work with more elegant and fine lines, and the mapping of places could be more precise. Corrections and additions could be made easily, and the metal plates could produce large maps. The woodcut process offered different advantages: it required just a common printing press, the woodblock accommodated type, and many copies could be pulled from a single woodblock.[3]

For the Ulm edition of Ptolemy's *Geography*, which used the woodcut medium, Holle hired a woodcutter to transform the painted originals of Nicolò Germano, who had created the maps for Borso d'Este's lavish manuscript edition of Ptolemy's *Geography* (plate 3). The woodcutter signed his *mappamundi*: "Engraved by Johannes Schnitzer, woodcutter from Armszheim." Holle then printed the entire *Geography* from the typeset woodblocks, some with the hand-cut maps by Schnitzer. The result is, according to map historian Rodney Shirley, "more crudely executed," but the "bold Germanic style and gothic lettering have a distinctive decorative appeal." Whether using copperplate engraving, such as in Italy, or the woodblock, which was preferred in Germany, printers became familiar with the process of printing maps.[4]

Maps appeared in other kinds of printed books, even though, in the words of historian Andrew Pettegree, "the world of books could be cruel" to the early printers, many of whom (including Gutenberg) were ruined. Among printers who went bankrupt was Holle, who lost his business shortly after printing the Ulm edition of Ptolemy's *Geography*. In 1493 Hartmann Schedel released the first edition of the massive *Liber Chronicarum* (*Nuremberg Chronicle*); a German edition appeared later in the same year. This printed book contained hundreds of woodcut illustrations, among them a world map derived from Ptolemy, a map of Germany, and many maplike views of cities.[5]

Because of the uncertainly of their trade, printers sought out materials besides books that could be printed and sold for profit. Short books, pamphlets, and illustrated bibles had proved to be good financial investments, and woodcut illustrations added value to them. Broadsides, which combined a woodcut illustration with several lines of text, proved worthwhile for printers. Printers hired artists to design images, and another artisan—the *Formshneider*—to cut the blocks. Woodcut maps—printed separately and independently of books—date from at least 1475 and flourished for one hundred years. The same printers who printed illustrated books and broadsides also produced woodcut maps for sale.[6]

These printed maps, however, did not include the emerging new view of the Atlantic that was visible on the charts used by mariners and merchants. On the *mappaemundi* in editions of Ptolemy's *Geography*, the Atlantic still appeared as a marginal ocean. It was precisely this omission that Martin Waldseemüller, Matthias Ringmann, and Walther (or Gualtier) Lud sought to correct through the publication of a book, a printed map, and a small map. All were humanists living under the patronage of Duke René of Lorraine in the Alpine village of St. Dié, not far from the city of Strasbourg. In addition to the world map, globe, and book, Waldseemüller, Ringmann, and Lud were working on an edition of Ptolemy's *Geography* that would include the original Greek text. They also planned to add new "modern" maps to the traditional maps conceptualized and described by Ptolemy.[7]

In 1507 a German Benedictine monk wrote to a friend that he had "purchased cheaply a beautiful small globe printed in Strasbourg as well as a large map of the world with islands and regions recently discovered by Amerigo Vespucci." The items selected are believed to be Martin Waldseemüller's 1507 large printed wall map of the world and the associated globe. Trithemius's purchase makes clear that Waldseemüller's map and globe could be purchased by anyone with the means to do so, and because they had been printed, they were relatively inexpensive. A book that described the map and globe was also printed and for sale.[8]

From his letter, it seems Trithemius acquired a finished globe. On the other hand, it is likely that he purchased just the printed sheets of the large world map. The globe measured about 11 cm in diameter (just under four and a half inches), and it could be held in one hand (figure 2.3). The map, however, came on twelve

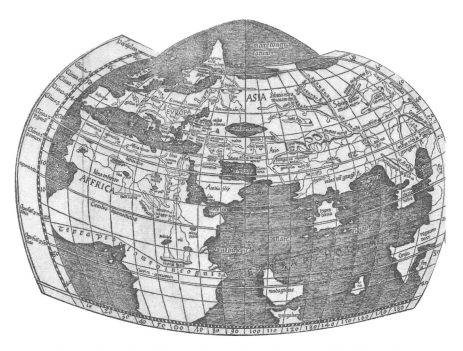

Fig. 4.1. Eastern (*above*) and Western (*right*) Hemispheres, 1507. Detail from Martin Waldseemüller, *Universalis Cosmographia*, 1507, Geography and Map Division, Library of Congress, Washington, DC.

separate sheets of paper. To be viewed in full, it needed to be properly sequenced, pasted on linen, mounted on a board, and hung on a wall. Once assembled, the map measured approximately 128 × 233 cm (about 8 × 4 feet)—too large for Trithemius to carry home. Even if overpowering in size, large maps in architectural spaces were popular in the sixteenth century. Whereas wealthy lords could hire artists to paint the walls of their palaces, a printed wall map offered the same instruction and delight at a fraction of the cost. Possibly, Trithemius's sheets were finished in color, making its aesthetics even more pleasing. Today the only surviving copy of this map is not colored (figure 2.2). Still its aesthetics are striking with the black ink contrasting sharply with the ivory-colored paper.[9]

Fully compiled, Waldseemüller's large map presented the viewer with a complex and detailed picture of the world and the important place of the Atlantic in it. Presiding at the top are Claudius Ptolemy and Amerigo Vespucci, and next to each are two smaller hemispheric maps. Ptolemy, identified as the "Cosmographer from Alexandria," holds a quadrant, an instrument he used to measure the altitude of the sun. In the adjacent hemispheric map (figure 4.1, *left*), Ptolemy looks out at the world that he had mapped: the Greek *oikoumene.* Waldseemüller has, however,

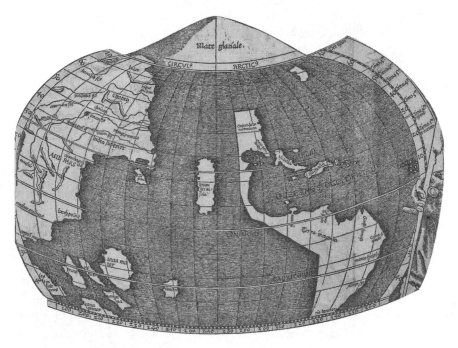

Fig. 4.1. (continued)

taken a small liberty—he shows the limits of Ptolemy's knowledge. A line that begins approximately 15° S bisects Africa, and under it Waldseemüller inserted a text
that states "Terra Ptholomeo Incognita," (Land Unknown to Ptolemy). This line
continues into the Indian Ocean, where again a text speaks: "Extra Ptholomeum"
(Outside Ptolemy). By adding geographic information unknown to Ptolemy, Waldseemüller signals that he is correcting Ptolemy with modern knowledge. Opposite
Ptolemy is Vespucci, the source of this new knowledge. Vespucci looks across at a
hemisphere that Ptolemy had not mapped (figure 4.1, *right*). Vespucci holds a compass divider, a key instrument used by chartmakers. By placing Vespucci opposite
to Ptolemy, Waldseemüller signals to viewers his belief that Vespucci is the modern
heir of the classical Greek authority. The portrait of Vespucci is even a bit cheeky:
Vespucci points with his pinky into the map, indicating what Ptolemy had not
known—the existence of another part of the world.[10]

Below Ptolemy, Vespucci, and their two hemispheres, Waldseemüller's map
bursts out in a wide, bulblike shape (figure 2.2). On the far left are the newly
discovered lands, with the name America inserted over South America. In the
top center left is Europe, dark with text, while Africa takes up most of the center
left. An immense Asia with an expansive Indian Ocean occupies nearly all of
the right. Tightly cut, curved lines present the oceans as dark and undulating.

White spaces left on the continents allow mountain ranges, rivers, lakes, and hundreds of place names to be visible. Boxes of descriptive text, as well as labels, appear everywhere.

The globe, which a viewer could hold, might have been painted. Trithemius describes the globe he purchased as *pulchra*—beautiful—which is suggestive of its having been finely finished. On the globe, the major parallels are all marked: the equator at the center, the Arctic and Antarctic Circles at the North and South Poles, the Tropics of Cancer and Capricorn. Lines of latitude are presented in parallels every ten degrees above and below the equator, and the longitude meridians are similarly placed in ten-degree segments. The major oceans are labeled as the "Oceanus Occidental" (Western Ocean), the "Oceanus Orientalis" (Eastern Ocean), and the "Mare Indicum" (Indian Sea). Their waters all connect. Africa and Asia appear as the largest continents. America is a thinly shaped island surrounded by the Oceanus Occidental (figure 2.3).[11]

It is impossible to know how Trithemius, or any other viewer of Waldseemüller's large wall map of the world and of the globe, understood them. However, it is possible to see how Waldseemüller sought to shape their interpretation through the book published at the same time. According to its long title, it was intended to be an introduction to the principles of astronomy and geometry, the four voyages made by Amerigo Vespucci, and the two maps—one *in solido* (the globe) and the other *plano* (the flat world map). Two dedications commit the entire project (book, large world wall map, and globe) to Maximilian I, who would be crowned Holy Roman Emperor the very next year. Nevertheless, in one dedication Waldseemüller reveals the larger intended audience. After wishing Maximilian good fortune, Waldseemüller makes the case that it is equally valuable to learn about the world through books as it is through travel. Waldseemüller explains that he has included material from the books of Ptolemy and has added further information from the four voyages made by Americo Vespucci. Waldseemüller then states that, in addition, he has made a world map and a globe "for scholars."[12]

Moreover, on the large world map, Waldseemüller used text extensively to ensure that viewers understood his "Fourth Part of the World"—the Oceanus Occidental (Western Ocean, i.e., the Atlantic) and the newly discovered islands in it (i.e., the Americas). On the large wall map, Waldseemüller placed labels where a viewer might expect to find them—such as next to cities or over oceans and continents. In addition, in descriptive texts he offered explanations of specific geographic places. A title runs across the bottom, and a formal introduction to the map appears in four boxes of text placed at the four corners of the map. The title and the explanatory texts speak directly to the map viewer, and they seek to orient the interpretation of the map.

Cultural historian Christian Jacob differentiates between the "silent map" and the "written map," and this observation helps decode how the Atlantic appeared on Waldseemüller's maps of 1507. On a "written map," Jacob explains, "the drawing is erased under a flood of words . . . writing invades geographic forms." Waldseemüller's large world map clearly falls into this category. On the other hand, Waldseemüller's globe has very little text and is closer to, if not an example of, Jacob's "silent map." The two hemispheric maps—next to Ptolemy and Vespucci at the top of the large wall map—fall in between: Ptolemy's hemisphere, tilting toward the "written," and Vespucci's Atlantic (and its newly discovered lands), toward the "silent." In Jacob's view, the "silent map delivers its message only if its reader actualizes the information it contains," whereas the written map—"saturated with information"—readily "answers a certain number of questions."[13]

Waldseemüller's large wall map is heavy with text for the known world—Europe, Africa, and Asia—which was also the world familiar to the potential viewers. For the map viewers to understand the new information that he has incorporated into the map, Waldseemüller uses different kinds of texts, which, following Jacob's insights about the "written map," play different roles in shaping the viewer's understanding of the map. The title is clearly significant—as Jacob writes, "the title is probably what most directly determines the perception and comprehension of the document." Waldseemüller's title clearly states what the message of the map is: *Universalis Cosmographia Secundum Ptholomaei Traditionem et Americi Vespucii Alioru[m]que Lustrationes* (A Complete Cosmography following the Ptolemaic Tradition and the Travels of Amerigo Vespucci and others). With this title, Waldseemüller addresses the map viewers directly, and he tells them what to expect. From Trithemius's letter, it is clear that he got this message—for he wrote his friend that the map he had purchased included new information drawn from Vespucci's voyages.[14]

Waldseemüller also used legends to explain the importance of what he called the "Fourth Part of the World," that is, America. As Jacob notes, a map's legend "expresses what cannot be represented or what in one way or another requires a verbal translation." Where legends are placed is also important for when the legend is removed from the map itself to a marginal inset, Jacob writes, "it leaves a place open for thematic developments, debates, or more structured narratives." At the four corners of Waldseemüller's wall map are such boxes, with texts addressed to the map's viewers. Their short paragraphs orient the viewer and provide background on how the map was made and how it should be read. In the top-left box, the text begins poetically by stating that a land once only "imagined" had been "discovered." Its discoverers were Columbus, "captain of the King of Castile," and Vespucci, and "the land extends about 19 degrees [roughly] beyond the Tropic of Capricorn toward the Antarctic Pole." The top right box provides Waldseemüller's rationale as a cosmographer: "It has seemed best to put down the discoveries of the ancients, and to add what has since been discovered by moderns." The lower-right box reiterates the importance of the new knowledge contained in the map: "Many things remained unknown to them [the ancients] in no slight degree; for instance, in the west, America, named after its discoverer, which is now known to be a fourth part of the world." The text concludes with a plea: "This one request we have to make, that those who are inexperienced and unacquainted with cosmography shall not condemn all this before they have learned what will surely be clearer to them later on, when they have come to understand it." The lower-left box cites the islands and mainlands discovered on four voyages made between 1497 and 1504—two under the sponsorship of Castile and two by the king of Portugal.[15]

The legends on the large wall map orient the viewer to what is new and even controversial in Waldseemüller's depiction of the world. These issues are more fully addressed in the book—*Cosmographiae Introductio*—published at the same time. In part 1, chapter 9, a reader encounters the fact that the fourth part of the world is inhabited but that it is different because of its isolation. Europe, Africa, and Asia are continents, connected to each other, but the newly discovered land is an island, completely surrounded by the sea. The reader learns that this newly discovered land or island in the fourth part of the world has been named "America," after Amerigo Vespucci, whom the authors credit with its discovery.[16]

Part 2 of *Cosmographiae Introductio* presents a Latin translation of a letter, written by Amerigo Vespucci and addressed to Piero di Tommaso Soderini, the *Gonfalonier di Giustizia* (Gonfalonier of Justice—an important official) of the Florentine Republic. The letter describes the four voyages, referred to in the lower left text box on Waldseemüller's large world map—two under the Spanish flag and two under the Portuguese—that Vespucci claimed to have made. In voyage one, Vespucci describes the Caribbean; in voyage two, he relates crossing the equator and sailing south along the northern coast of South America before returning to the Caribbean; on voyage three, he claims to have sailed all along the coastline of South America, passing even beyond the Tropic of Capricorn. In voyage four, Vespucci returned to the Bay of All Saints, which he stated he had visited on the earlier voyage (to Brazil), then sailed again along the coast of South America, reaching 18° south of the equator where he built a fort, leaving behind twenty-four men with provisions for six months. Thus, according to this letter Vespucci departed on his first voyage from Cadiz in May 1497; on his second from Cadiz in May 1499; on his third from Lisbon in May 1501; and on his fourth from Lisbon in May 1503.[17]

The humanists at St. Dié believed they had found a text that adequately described the "fourth part of the world"—a highly populated land across the ocean from Europe and Africa—written by a credible eyewitness. Yet Vespucci's letter, known today as the *Lettera*, or as the *Soderini Letter*, or as the *Quattro Viaggi* (*Four Voyages*), is considered by scholars to be Vespucci's least credible. Questioning whether Vespucci actually made all four voyages, scholars note his exaggerated and repeated claims to be a highly experienced navigator. The *Lettera* contains vivid descriptions of cannibalism, which many believe were added either by Vespucci to embellish the text or by the editor to make the printed text more appealing to readers.[18]

On the globe, the large world map, and the hemispheric maps at the top of the *Universalis Cosmographia*, Waldseemüller included this fourth part of the world (figure 4.2). The oceans are large and bold on all three of the maps. On the large wall map, the Atlantic is especially prominent, and Waldseemüller labels it the "Oceanus Occidentalis" (Western Ocean). The ocean off the coast of Northwest Africa on the large wall map and on the hemispheric map next to Ptolemy has an additional label: "Athlanticum" (Atlantic). This specific label is explained in the *Cosmographiae Introductio*, where it is noted that "Athlantico" refers to waters lying to the west of Europe and Africa and that Europe is bounded on the west by the "mari Athlantico" (Atlantic Sea), as is Africa.[19]

The rationale behind the label "America," placed on both the globe and the wall map, is also given an explanation in the *Cosmographiae Introductio*: "And the fourth part of the earth, which, because Amerigo discovered it, we may call Amerige, the land of Amerigo, so to speak, or America." Later, more elaboration is offered, but almost as an aside: "Because it is well known that Europe and Asia were named after women, I can see no reason why anyone would have good reason to object to calling this fourth part Amerige, the land of Amerigo, or America, after the man of great ability who discovered it."[20]

Waldseemüller placed the fewest labels on the globe—"America" for the land and "Oceanus Occidental" (Western Ocean) for the water (figure 4.2, *center*). On the hemispheric map (the one next to Vespucci at the top of the wall map), there is more text: Waldseemüller added labels for the islands of Cuba ("Isabella") and Hispaniola ("Spagnolla") and for two places in North America. Over South America, he inscribed "Terra incognita" (unknown land). Waldseemüller then gives seven place names for South America that refer to geographic features, such as capes, rivers, and a mountain (figure 4.2, *left*). On the huge world map, there is considerably more text, even as the most prominent emphasize that the two enormous land masses are "ultra incognita" (very unknown). There is little characterization of the landscape, except for mountain ranges and rivers. A bird is

Fig. 4.2. Waldseemüller's Fourth Part of the World on Three Maps, 1507. Details from Martin Waldseemüller, *Universalis Cosmographia,* Geography and Map Division, Library of Congress, Washington, DC (*left and right*) and re-created globe from Martin Waldseemüller, *The 1507 globular map of the world,* courtesy of the James Ford Bell Library, University of Minnesota, Minneapolis (*center*).

the only illustration. For labels, Waldseemüller used "Parias" for the northern land and "America" for the southern. Along the eastern coastline of America, Waldseemüller inscribed toponyms taken from charts, inserted Spanish and Portuguese flags to mark places visited, and used text boxes to describe the history of discoveries. One such text box appears off the coast of Brazil, next to a ship, and relates its discovery in 1500 by Cabral (figure 4.2, *right*).

In the *Cosmographiae Introductio* America is described primarily as an "island" and only occasionally as a "land." In part 1 chapter 9, the authors write: "The earth is now known to be divided into four parts. The first three of these are continents, but the fourth part is an island because it has been found to be completely surrounded on all of its sides by sea." This is repeated in the lower left text box on the *Universalis Cosmographia,* which reads in part: "A general delineation of the various lands and islands, including some of which the ancients make no mention, discovered lately between 1497 and 1505."[21]

The *Cosmographiae Introductio* also reveals that Waldseemüller mapped the globe differently from the large wall map. Regarding the globe, the book states that "we have followed the description of Amerigo." This suggests that Vespucci's *Lettera* (as well as the *Mundus Novus*) served as important sources. The *Mundus Novus* letter described one transatlantic voyage (to Brazil), whereas the *Lettera* narrates four. Taken together, these voyages describe a huge area extending from the Caribbean to South America beyond the Tropic of Capricorn. These enormous distances covered much of the area known to Spanish and Portuguese sailors, yet Vespucci's *Lettera* and *Mundus Novus* provide little detailed geographic information. This makes clear that Waldseemüller turned to charts to fill in the details that are absent from the globe but visible on the large wall map. As explained in the *Cosmographiae Introductio*, "For us, therefore, it has been necessary, as Ptolemy himself suggests, to place more faith in the information gathered in our times. We have on our map therefore followed Ptolemy, added new lands and some other things."[22]

This comment would explain why the placement of the major parallels on the large wall are markedly different from the globe and the hemispheric maps. On the large world map, the equator and the Tropics of Cancer and Capricorn are not placed correctly with respect to the toponyms for either South America or Africa. For example, Waldseemüller has the Tropic of Capricorn running through the Bay of All Saints ("Abbatia omnū Sanctorum"), when in fact the Tropic should be positioned much farther south, next to Rio de Janeiro ("S. Sebastiam"). Similarly, the equator and the Tropic of Capricorn are incorrectly placed in Africa. The equator should run through São Tomé, yet Waldseemüller has positioned "S. Thome" (São Tomé) above the Tropic of Capricorn. On the globe, these parallels are more correctly placed for South America, even as the equator is still too far north in Africa, and the Tropic of Cancer too far north in North America.

Waldseemüller does not name his sources; however, scholars suggest the maps and charts he had before him. Clearly, Waldseemüller valued Ptolemy, and he probably knew Holle's edition of Ptolemy's *Geography* printed at Ulm. He and Ringmann also sought out other editions of Ptolemy, primarily for the edition of *Geography* that the group in St. Dié intended to publish, and these would have also influenced the mapping of the large wall map. Ptolemy emphasized the importance of modern knowledge, and Waldseemüller sought out modern *mappaemundi* as well as current charts. Chet Van Duzer has shown convincingly that there are striking similarities between Waldseemüller's *Universalis Cosmographia* and the large map of the world by Henricus Martellus of circa 1491. After using hyperspectral imaging techniques to view the illegible legends on the world map, Van Duzer concluded that Waldseemüller did not directly copy Martellus so much as he picked and chose what he wanted to incorporate. Van Duzer's study of legends reveals that Waldseemüller borrowed extensively from Martellus for his mapping of Asia and Africa, except when he had better information from charts. Waldseemüller relied on *cartas marinas* (charts) for the Atlantic and Indian Oceans. Scholars have long argued that he had access to the world chart signed by Nicolay de Caverio (plate 7).[23]

To make a printed map required a collaboration between the mapmaker and the printer, and this collaboration would influence how the Atlantic appeared on Waldseemüller's maps. As map historian Arthur Robinson perceptively notes, for a cartographer to take advantage of the opportunities offered by print—rapid production of numerous copies of the same map—"the cartographer had to submit to the controls imposed by the printing processes." At the same time, the cartographer placed demands on the printer regarding size, consistency, and accuracy. In contrast, Robinson reflects, "the manuscript mapmaker is his own master." As noted in the previous chapter, once chartmakers had laid out the underlying geometry, they worked as artists, using quills filled with ink and brushes daubed with paint. Working alone, the chartmaker could perform every step in the process, and even if a busy chartmaker had multiple "hands" working on a chart— one lettering in the labels, another painting the illuminations—the style of the chart, and especially distinctive compass roses, reflected the aesthetics of the chartmaker. Similarly, when artists copied a chart or created their own maps from Ptolemy's instructions, or used a grid to plot a view of a city, their finished maps carried their own artistic selections. Such decisions as the color picked to indicate water, or the width of the lines drawn to represent mountains, shaped not only the geographic information but the beauty of the map. In the case of a printed map, however, the printer hired specialized artisans—the designers, woodcutters, typesetters, pressmen, and colorists. These artisans could influence not only the "look" of a printed map but also the shape of geographic features, such as an ocean. In considering how the Atlantic appeared on Waldseemüller's printed map and globe, one must consider both the mapmaker and the unnamed artisans.[24]

As soon as it was decided to produce a printed world map and globe, Waldseemüller had to have been in conversation with printers. Martellus's large printed world map might have been a model, or even the inspiration, for printing the *Universalis Comsographia*. Waldseemüller might also have known of and personally seen the large and detailed woodcut *View of Venice*, which was printed in 1500 and sponsored by Anton Kolb, a German merchant from Nuremberg. Jacopo de' Barbari designed the huge view—almost 4 × 4 m (about 13 × 13 feet)—that was printed on six oversized sheets of paper (each sheet measured approximately 70 × 100 cm). The multiple-sheet woodcut presents in great detail the streets, buildings, neighborhoods, canals, and boats of Venice, all from an aerial perspective. Kolb protected his rights by receiving permission from the Venetian Senate to sell and export the map for four years. The finished map sold for three ducats, and its intended clientele was educated and prosperous.[25]

No sources describe how Waldseemüller interacted with the printer or with the specialized artisans, and there is no certainty even about where it was printed. The book and the globe were printed on a small press at St. Dié, but the large wall map presented greater challenges. Printing the globe gores was relatively simple: once Waldseemüller created the map in the gore format, the design was transferred onto the woodblock, cut by the *Formschneider*, and printed on single sheets of paper. The book was more complicated, but it too was also fairly straightforward, primarily setting type and inserting the woodcut illustrations. The large map of the world, however, was a very different proposition. To make the map, twelve sheets were printed from twelve separately cut woodblocks. The printer would need specialized artisans—typesetters, woodcutters, and pressmen—who were not likely to live in St. Dié. This has led many scholars to suggest that the large map was printed in Strasbourg.[26]

Before the collaboration with the printer and the artisans began, Waldseemüller first had to create the map, and we can assume that he worked from Ptolemy's instructions, modifying them as necessary. The first step, therefore, was laying out the projection, and this would be no easy task. As John Hessler observes, "Among the many technical and theoretical problems that Waldseemüller faced in the construction of his map, one of the least trivial mathematically was the problem of projection." Ptolemy provided explicit instructions, beginning with the size of the surface and the proportions of the sides, and these steps required basic geometry, which Waldseemüller, using a compass and ruler, would have no trouble following. His problem was that Ptolemy's known world was much smaller than the world he intended to map. Waldseemüller knew that Ptolemy's world was too small and that the newly discovered "Fourth Part of the World" would not fit. According to Hessler, "Dealing with a greatly enlarged earth, compared with the Ptolemaic models at his disposal, Waldseemüller modified Ptolemy's second conic projection." As is true for any map, the projection adopted will shape how continents and oceans appear. Ptolemy's spherical projection favored the Northern Hemisphere, but Waldseemüller needed to extend it to the west and to the south. The result, according to Hessler, was that Waldseemüller "distorted the shape of the new continents because they were forced to the far western portion of the map and hence greatly elongated."[27]

Ptolemy recommended an order to drawing the map: the hand should move from left to right and from top to bottom. He advised drawing the northern places before the southern and the western before the eastern. For Ptolemy's known world, this meant beginning the map with Europe, then drawing North Africa, followed by Asia. Although we cannot be certain what order Waldseemüller followed, it seems unlikely that he would begin at the top left, with North America, the most uncertain, least understood part of the map. Instead, he probably began with Europe and followed Ptolemy's recommended progression—Europe to Africa to Asia—until he had all of the classical knowledge entered. Then using his additional sources, he would add in, as best he could, the locations that appeared on charts for South Central Africa and the Americas.[28]

Most likely, and this is an educated guess, Waldseemüller worked with a quill pen on large sheets of paper. No drafts of the map survive. That he might have worked on paper can be inferred from a remark made by a printer in Strasbourg. Art historians David Landau and Peter Parshall describe a letter written by Johann Grüninger to Willibald Pirckheimer, who wished to print a world map. Grüninger reported that the map was "nearly drafted on paper and could soon be put onto the block for cutting." Arthur Robinson calls this the "map work-sheet" or the "manuscript drawing made by the cartographer."[29]

While Waldseemüller was finishing the map worksheet, the woodcutters were preparing the twelve woodblocks. The individual blocks were composed of separate planks of wood that were glued and reinforced by butterfly joints or metal cleats, and possibly even with crosspieces screwed into the backside. Once assembled, the face of each block would have been sanded smooth and possibly a thin white coat of paint applied.[30]

When Waldseemüller finished the map worksheet, and when the woodblocks were ready, it was time to transfer the map onto the woodblocks. Because in the woodcut process gouged-out grooves appeared white on the finished printed sheet, while the black ink adheres to the raised ridges left behind, the original map had to be completely reversed for the woodcutter. Various methods might have been used, from carboning (pressing down on the design and tracing it over a thin film of carbon), incising and pouncing (pricking tiny holes in the paper and lightly smacking the paper with a pouch full of colored chalk), oiling the sheets to make them transparent, turning them over, and gluing them onto the block. Landau and Parshall suggest that detailed designs may have been drawn directly onto the woodblocks. However it was done, this was highly skilled work, even an artistic specialty.[31]

In this process, some of Waldseemüller's original design most likely had to be modified into strokes that the woodcutter could easily and effectively cut. Mountain ranges in Europe, Africa, and Asia became thick, sticklike forms. Described as a "caterpillar style" by R. A. Skelton, two lines enclosed the range, and closely cut, repeating lines created shading to suggest height. Two undulating lines, indicative of riverbanks, represented rivers. An interesting difference can be seen in the mountain ranges in the Americas as compared to those of Europe, Asia, and Africa. The mountains of the three known continents are designated by a thick line that defines the form of the range, while short, thinner lines drawn next to each other suggest height. The map viewer can easily read the image and understand it to mean mountain ranges. The mountains drawn in the Americas, on the other hand, are presented in a more abstract way. They are cut to emphasize depth, height, and three-dimensionality. Possibly another designer and cutter worked on this section of the map, or the greater space offered by the largely blank (and unknown) continents presented the designer and woodcutter with the opportunity to experiment with new forms (figure 4.3).[32]

Artisans, and especially the woodcutter, probably had a voice in how the oceans would be shown. The oceans on the *Universalis Cosmographia* are cut with lines that closely follow each other in gradual curves, enhanced here and there with the crests. When printed, and possibly tinted with blue, the effect would convey movement and depth. The wave crests do not appear on the globe or on the hemispheric map next to Vespucci, but one does appear in the Indian Ocean in the hemispheric map next to Ptolemy. Otherwise, the oceans are cut similarly on the hemispheric maps, the large world map, and the globe. The cutting of the oceans on Waldseemüller's three maps printed in 1507 is distinctive but not completely original. Woodcutters had already developed techniques to render oceans. On the world map in Hartmann Schedel's *Nuremberg Chronicle*, the woodcutter used a similar style of closely cut, wavy lines to depict the Indian Ocean (figure 4.4, *top left*). In the *View of Venice,* de' Barbari renders water in a variety of ways, with closely cut straight lines employed to suggest a placidity and distance (figure 4.4, *lower right*), but in the southwest corner, the cutting is done to suggest waves (figure 4.4, *lower left*). On Waldseemüller's *Universalis Cosmographia*, the South Atlantic appears full of movement with cresting waves (figure 4.4, *top right*).

The border of the *Universalis Cosmographia* contrasts with the map and reveals a different style of woodcutting. It too was a collaboration with Waldseemüller but one where the artisans had more influence. Because of Waldseemüller's projection, which created a bulb- or bean-shaped view of the world, there is an empty space that surrounds his map. At the top are the portraits of Ptolemy and Vespucci, which we may assume that Waldseemüller requested, along with the hemispheric maps associated with each, which Waldseemüller would have created.

Fig. 4.3 Mountains on Waldseemüller's *Universalis Cosmographia.* Details from Martin Waldseemüller, *Universalis Cosmographia,* 1507, Geography and Map Division, Library of Congress, Washington, DC.

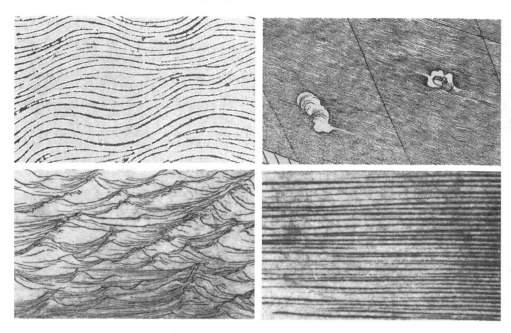

Fig. 4.4. Woodcutters' Techniques for Bodies of Water. Details from Hartmann Schedel, *Nuremberg Chronicle*, 1493 (*top left*); Martin Waldseemüller, *Universalis Cosmographia*, 1507, Geography and Map Division, Library of Congress, Washington, DC (*top right*); Jacob d'Barbari, *View of Venice*, 1500, Cleveland Museum of Art (*bottom left and bottom right*).

It fell to the skill of the designer and woodcutter to execute the portraits of Ptolemy and Vespucci. Both stand, their bodies turned at an angle, with a bank of clouds behind them. Ptolemy wears a turban wound elaborately around his head, its sash resting on his cloak, which falls in folds around his arms. His face is aged, but his eyes are still clear; his beard is thick. Vespucci wears a cap, and his hair hangs in long curling locks, as do the hairs of his beard. The folds of the cloth drape around his arm. Below each are windheads: Circius next to Ptolemy and Aquilo next to Vespucci. Again, the skill of the designer and woodcutter is visible. Circius (figure 4.5, *left*) and Aquilo (figure 4.5, *right*) have curling locks of hair, their cheeks are puffy and full, and careful shading gives their heads and faces depth. A third windhead, Septentrio, emerges from the space where the two hemispheric maps meet, and one half of Septentrio falls on Ptolemy's side and the other on Vespucci's (figure 4.5, *center*). Each of the windheads blows wind, symbolized by long thin blades that emerge from its mouth and extend into the map.

Ringmann and Waldseemüller devote a whole chapter in the *Cosmographiae Introductio* to the twelve winds, making the placement of windheads—as well as

Fig. 4.5. Three Windheads at the Top of Waldseemüller's *Universalis Cosmographia.* Details from Martin Waldseemüller, *Universalis Cosmographia*, 1507, Geography and Map Division, Library of Congress, Washington, DC.

their execution by the artisans—a key step in the map's design that would have been important to Waldseemüller. Beginning with the philosophers, Ringmann and Waldseemüller define wind as "an exhalation, warm and dry, moving laterally around the earth." Then citing the Latin poets Ovid and Virgil, as well as a contemporary German, they differentiate between the winds. For example: "The north wind (Aquilo), by reason of the severity of its cold, freezes the waters." Windheads had appeared on the borders of Ptolemaic maps, both in manuscript and in print, especially surrounding the world maps. From this prior practice, as well as the attention paid to the winds in the *Cosmographiae Introductio*, we can assume that Waldseemüller insisted that each of the twelve windheads be distinct (as befitted their known natures), be positioned correctly, and be named in Latin. However, once Waldseemüller had indicated what windhead went where, the drawing and cutting of the windheads fell to the designer and the woodcutter. The artists set each against a background of closely drawn parallel lines—suggesting the sky— and billowing clouds. The curved lines enhanced with hatching created fat cheeks and pursed lips, while long tapering lines shaped hanks of hair with flowing curls.[33]

The distinctive clouds that surround the windheads on the border of Waldseemüller's map also gave the artists an opportunity to use hatching techniques and long curving strokes, the first to suggest depth and the second to convey movement. On the map itself, this technique is visible only in the cutting of the oceans with their cresting waves, in the different style used to depict the mountains in the Americas, and in the execution of a few illustrations, such as the ship off the coast of Brazil or the elephant in Africa. The cutting of the Europe, Africa, and

Asia is more restrained and more suggestive of what Waldseemüller's original map worksheet might have looked like. In contrast, the oceans and the border have a wilder, more intense look.

The design and execution of the border with its windheads and clouds, as well as the portraits of Ptolemy and Vespucci, are suggestive of new ways of cutting blocks introduced by Albrecht Dürer. These techniques had, in the last years of the fifteenth and early years of the sixteenth century, revolutionized the medium of the woodcut. As an apprentice in 1492, Dürer worked on the *Nuremberg Chronicle* with its many woodcut illustrations and city views. In 1498 Dürer designed, cut, and printed in Nuremberg the short, folio-sized book with fourteen full-page woodcuts, titled *Apocalypse*. Dürer reduced the book of Revelation into fourteen scenes of dramatic, momentous, apocalyptic action. He separated these scenes from the biblical text, which appears on the facing page, so that the viewer could either read the text or contemplate the image. Art historians recognize the woodcuts as especially innovative and transformative. Dürer's new style created what art historian Erwin Panofsky calls a "dynamic calligraphy" that emulated copper-plate engravings. Dürer's lines of different lengths and widths moved in curves. This new technique allowed the artist to work with positive and negative space, opening up new possibilities. The artist had a scale of chiaroscuro values that allowed black lines to suggest depth and shade, whereas places left white would appear brilliant in contrast. The woodcut, writes Panofsky, "began to glitter and glow with varying degrees of luminosity."[34]

It has long been suggested that Dürer, or someone influenced by him, might have worked on the Waldseemüller's second world map, the *Carta Marina*. Dürer spent some time in Strasbourg in 1490s, and he was certainly familiar with printed maps. He probably knew Jacopo de' Barbari's *View of Venice*, as Dürer had traveled to Venice. There is no evidence that Dürer was in Strasbourg during the time the *Universalis Cosmographia* would have been made. In 1506, he was in Venice, returning to Nuremberg in 1507. Still, the artist who designed and the woodcutter who executed the border, the windheads, and the portraits of Ptolemy and Vespucci had been influenced by Dürer's woodcutting techniques. There were excellent woodcutters in Strasbourg who might have worked on the map, such as Hans Baldung Grien, Hans Wechtlin, Erhard Schlitzoc, and Hans Franck—all of whom worked for Johann Grüninger, one of the possible printers for the *Cosmographia Universalis*.[35]

What does the elaborate border do for Waldseemüller's map? It was modern in its time, for the new woodcutting techniques presented clouds and windheads in dramatic new ways. At the same time, what it depicted—the celestial sphere with its winds with their classical names—was familiar and known. The combination of the new with the old repeats in the map where the new, fourth part of the world is integrated into the old known world. Maps that once showed windheads in association with Western Europe and Africa now blow their winds over new islands and mainlands in the Atlantic. The air coming from the windheads' mouths, like knife blades, cut into the map, thereby drawing the viewers in and asking them to look carefully at the map. The new woodcutting techniques used in the border frame and integrate the new geographical knowledge. Christian Jacob suggests that the border is "a transitional space for the gaze and for thinking" and that when the border has a turbulent mix of winds and clouds, this represents "the machinery of the cosmos"—a celestial space apart from the human world. The windheads and the clouds on Waldseemüller's wall map invite the map viewer to reflect on the cosmos and to see how it relates to a new "Fourth Part of the World."[36]

Artisans tasked with transferring Waldseemüller's text onto the map worked with type and knife. Some of the text and many of the legends were set with moveable type and later inserted into the block, but many of the toponyms and labels had to be cut by hand. Each name cut by hand was reversed, and each letter had to be sequenced correctly in order that words not be misspelled or letters accidentally reversed. This task required great skill as well as careful focus.[37]

Five different typefaces have been identified on the map: two Roman and three Gothic. Renaissance humanists preferred Roman type, and by 1500 it was accepted everywhere for printing in Latin. Gothic, or black letter, resembled handwriting, and it remained the type of choice for printing in German. Whether using Roman or Gothic type, the setting of the text was the same. The compositor worked with an original that showed the text to be set, selecting one by one the upper- or lowercase letters for each word and positioning them into the composing stick. Also a highly skilled artisan, the compositor not only had to select the correct letters in upper or lower case and space them evenly using metal inserts but had to place everything into the stick backwards so that, when printed, the text would read from left to right. Each of the legends on Waldseemüller's map would have been encased into an iron or wood frame, which was then inserted into holes cut into the woodblock.[38]

Printing began when each block was finished. First, a trial copy or "proof" was made and presumably returned to Waldseemüller for corrections. Once all sheets had been read and any errors corrected on the woodblock, the press was set up for production. In advance of this, the printer had ordered the paper. The paper had to be prepared a day ahead, because it had to be damp to absorb the ink. When it was time to print, two pressmen worked together. A key first step was "registering" the paper, that is, making sure that the woodblock would be correctly centered on the page. One placed each dampened sheet carefully, taking care to follow the guides set up during the registration process. When the block was evenly inked, the first pressman lowered the tympan frame onto the bed and slid it under the press, while the second pressman pulled the bar to screw the press closed. Once the ink was judged to be sufficiently transferred onto the paper, the screw was released, the carriage pulled out, and the printed paper carefully pulled off and hung or laid out to dry.[39]

Waldseemüller's map would be printed with black ink on white paper, yet Waldseemüller and the designer allowed for future hand-coloring. That Waldseemüller imagined his map in color can be deduced from his discussion of several of the symbols that were to appear in the map. For example, the boundaries of Asia Minor were marked with "the saffron-colored" cross of the Turkish sultan, whereas a "red cross" symbolized the lands in Africa of the legendary ruler Prester John. The only surviving copy of Waldseemüller's map does not show any original coloring, and it is uncertain whether the printed sheets were finished in color. But clearly, if a patron requested it, or if the printer deemed it profitable, an illuminist could color the printed map sheets. Coloring the black and white map changed its appearance and readability. Brown applied over the stick and caterpillarlike cuts enhanced the depiction of mountain ranges, while blue painted between the riverbanks, outlined in black, made them more visible. By adding blue to the black-inked lines that already symbolized the seas and oceans, the colorist emphasized undulating movement and depth. The curls of waves, especially visible in the still unnamed Pacific Ocean, could be left white, and the effect would suggest the power of the vast, unexplored ocean. With the added color, the map appeared more impressive, thereby increasing its value, which would bring in extra income for the printer. For this reason alone, it seems possible that at least some copies of the map were once colored.[40]

Despite the key role played in the creation of the *Universalis Cosmographia,* the printer cannot be established with complete certainty. Historians commonly use typefaces and watermarks in the paper to detect printers and the dates of printings, however, the sole surviving copy of Waldseemüller's map has several odd features that make such an identification difficult. The first concerns the Roman and Gothic typefaces. According to Elizabeth Harris, the Roman typeface was probably the official type for the map, and it was the same type used to set and print the accompanying book—the *Cosmographiae Introductio.* Harris describes this type as "light, open and well balanced." Competing with this elegant type however is the Gothic type, which Harris describes as "small, dark, and dense." All the text on the large world map that is set in the Roman type, which includes the title along the bottom, is "consistently clean and competent" in Harris's estimation. In contrast to the Roman type, the text set in the Gothic type, Harris writes, "is sometimes so full of errors that the label is meaningless." Given the scholarly reputations of Lud, Ringmann, and Waldseemüller, these errors in the Gothic type are inexplicable. Also strange is the fact that some legends combine both Roman and Gothic types. Given the attention paid to the design of the map, it seems peculiar that the compositor would not only mix two typefaces but, even more unlikely, use different sizes of type in the same legend.[41]

Johann Grüninger is one possible printer for the large wall map. His shop in Strasbourg was large, and he employed skilled designers, woodcutters, and typesetters. He was well set up to print many copies of the map and to ship them out to Frankfurt for the book fair. The other possible printer of the map is Johann Schott. Schott followed in the footsteps of his father, Martin Schott, and became a printer, inheriting his father's shop in 1500. Previously Johann had studied at the same university as Waldseemüller; he was thus a humanist and certainly read Latin. In 1513, he printed the Ptolemy *Geography* that is attributed to Waldseemüller, Ringmann, and Lud, and he used the very same paper with the crown watermark that was used for Waldseemüller's *Universalis Cosmographia* and for Waldseemüller's later world map, the *Carta Marina*. Because the use of this paper is found only on these three exemplars, Hessler, Karnes, and Wanser speculate cautiously that Johann Schott printed the Waldseemüller world map of 1507.[42]

The dating of the sole surviving copy of the *Universalis Cosmographia* is also complicated. The accompanying book, the *Cosmographia Introductio*, carries the date of 1507, and Waldseemüller describes having printed the wall map and globe in a letter dated February 1508. However, the watermarks on the paper used to print the *Universalis Cosmographia* suggest a later date. Because watermarks are made from delicate pieces of wire fastened to the molds used to make sheets of paper, scholars use them to date and localize paper. The same paper with the same watermarks that was used to print the *Universalis Cosmographia* was also used to print Waldseemüller's second world map: the *Carta Marina* in 1516 and the edition of Ptolemy's *Geography* in Strasbourg in 1513. This evidence leads Hessler, Karnes and Wanser to conclude that the single remaining copy of Waldseemüller's 1507 world map, was printed between 1513 and 1516. When this printing of the world map was made, "the blocks were worse for the wear and had been altered in a few places," writes Elizabeth Harris. Possibly, some of the underlying woodblocks may have been damaged to the point where they could no longer hold the type. Such seems to be the case for the two frames that held two of the four explanatory text boxes. The top two text boxes appear as they must have been originally printed in 1507. The bottom two, however, are patched with pieces of paper that have been pasted over the map sheets. The two pieces of pasted paper were printed separately, presumably with the same text (but not in the same typeface) as on the originally designed map. One of the pieces of paper was a piece of scrap paper lying around the print shop. On its back is printing from a popular book on palm reading: *Ein schönes Buchlind der Kunst Chiromantia*, by Andreas Corvus, a book that was printed by Grüninger in Strasbourg in 1515.[43]

Thus the only copy of the *Universalis Cosmographia* that endures today is not the original map as designed and printed in 1507. The date of the printing of the scrap paper pasted into the lower-right block, the watermarks, the wear and tear on the blocks, and the odd dueling typefaces all confirm that the map had been printed in 1515 or after. It is possible that a special custom reprinting of the Waldseemüller 1507 world map was done expressly for the purpose of inserting each of the twelve sheets into a volume. This would have been commissioned by the Nuremberg humanist and globemaker Johannes Schöner, who bound all twelve sheets of Waldseemüller's world maps into a compilation of texts known as Schöner's *Sammelband*. Schöner undoubtedly saved these maps and texts for his own study, but John Hessler, who describes the compilation as "the most important collection of maps and geographical information to survive from the Renaissance," notes that Schöner consciously preserved them for future generations.[44]

Waldseemüller intended the huge *Universalis Cosmographia* and the small globe to make visible a formerly unknown part of the world. Central to that was the Western Ocean and the existence of a previously unknown land, or island, which he named America. A team of artisans working with a printer transformed Waldseemüller's maps from manuscript to print and enhanced them through highly skilled design and woodcutting. The elaborate border invited contemplation not only of the new "Fourth Part of the World" but also its place within the celestial sphere. The portraits of Ptolemy and Vespucci reinforced Waldseemüller's message that both old and new knowledge was needed to understand the world after 1500. Waldseemüller created the world map, the hemispheric maps, and the globe gores, but he could not execute every step in the making the printed maps. Nevertheless, the collaboration with the artists, artisans, and printer worked in his favor. The attractive globe and printed map sheets reached many viewers and conveyed the importance of the Atlantic Ocean and its newly discovered lands. Above the label "America" on the large map of the world appears a bird, perching on a scroll that reads "Rubei Psitaci" (Red parrot). This choice to use a parrot to signify newly discovered lands could have come from written texts, or have been suggested by the artisans, but it is more likely that it came from the charts Waldseemüller consulted. In addition to supplying new geographic information, world charts such as La Cosa's *Carta Universal* and *Carta del Cantino* introduced a new visual language to symbolize the Atlantic World, as we shall see in the next chapter.

CHAPTER FIVE

Parrots and Trees

How would the Americas be illustrated on charts and maps? There is no record of the first visual images used for the Americas until La Cosa's *Carta Universal* of 1500. In illustrating the Americas, La Cosa transferred known and easily interpreted imagery, such as flags and ships, that had appeared on many earlier charts. In addition, La Cosa entered the earliest known visual cartographic code that conveyed the nature of the Americas—the color green. Two years later, the chartmaker of *Carta del Cantino* presented two more images to describe Brazil: parrots and trees. These first visual codes conveyed meanings, and they continued to appear on subsequent charts. As images for the Americas migrated from chart to chart and from manuscript charts to printed maps, they began to characterize the place of the Americas in an Atlantic World.

Denis Wood identifies ten cartographic codes that make maps work; these codes create the framework, the conventions, and the rules that say "this means that." Among the most important is the iconic code, for "the map is an icon, a visual analogue of a geographic landscape . . . the product of a number of deliberate, repetitive, symbolic gestures, carefully arranged." Similarly, J. B. Harley describes the importance of the visual codes of maps by describing maps as "nonverbal sign systems" similar to paintings or prints; he emphasizes that maps have "a graphic language" that requires decoding. Because the viewer must be able to read the map, a mapmaker will favor those signs, symbols, and vignettes familiar to the maps' intended viewers. Yet, as Christian Jacob remarks, the cartographer "is also blessed with considerable freedom in presenting the whole of the drawing, in the choice of motifs, and finally, in the very execution of a graphic gesture."[1]

Juan de la Cosa painted the new lands in the western Atlantic a deep green on *Carta Universal* (1500). This is the first distinctive visual code for the Americas, and the choice of green most likely echoes the reports of thickly forested islands in the Caribbean, as described by Columbus and others. In the *Diario*, which is Bartolomé de las Casas's abstract of Columbus's log, "very green trees" are noted at the first landing on 11 October 1492. Trees continue to be noticed by Columbus and are described in Las Casas's extract as existing in "groves" with "thousands of kinds" that are "green," "beautiful," and "very large." When

Columbus arrived at Cuba, the *Diario* states that "he never saw such a beautiful thing, full of trees all surrounding the river, beautiful and green and different from ours, each with its own kind of flowers and fruits." La Cosa, who was on this voyage, would have been familiar with Columbus's first impressions, and may have had similar thoughts himself. The dark shade of green on *Carta Universal*, it would seem, conveys this first understanding of the nature of the new lands across the Atlantic. Moreover, the use of green obviates the need for specific information. La Cosa had reliable geographic information only for the islands of the Caribbean and the stretch of coast between the mouth of the Amazon River and the Golfo de Santa Marta (present day Colombia). By painting the land green, he did not need to enter any more geographic information; it was enough to depict them as suggestive of woodlands.[2]

Two years later, the color green repeats on *Carta del Cantino*; it is the dominant hue that depicts the Americas. Over South America, individual trees interrupt the green. Trees mark a new entry in the visual code for depicting the Americas. While green does imply woodlands on La Cosa's *Carta Universal*, the rendering of actual trees makes a more specific argument: this place has great forests. A third entry is the parrot. On *Carta del Cantino*, three enormous parrots, painted in vibrant hues of red, green, and blue, stand before a lush stand of trees.

No sources describe what a chartmaker was thinking when introducing a new visual code, and it is possible that the color green, parrots, and trees had appeared on earlier charts that have not survived to our day. Similarly, it cannot be ascertained if a chartmaker hired an illuminator to paint the imagery or how the "hand" of the chartmaker (or illuminator) might have influenced the rendering of a parrot or tree or even the particular shade of green. Nor is it possible to determine if a patron intervened and shaped the "look" of the chart by requesting (or even dictating) the illuminations. Although what these chartmakers, illuminators, and patrons intended is unknowable, the surviving maps themselves reveal that the color green, parrots, and trees defined a western land in the Atlantic on *Carta del Cantino* in 1502, and we can follow these codes as they repeated on Nicolay de Caverio's *Planisphère nautique* of 1506 (plate 7) and (with the exception of the color green) on Waldseemüller's *Universalis Cosmographia* of 1507 (figure 2.2). This movement from chart to chart and from manuscript chart to printed map demonstrates that chartmakers and mapmakers were alert to new visual codes and adopted them quickly. However, the images changed as they moved. The original specificity of the parrots and trees is simplified with each move, and the color green disappears on the sole surviving copy of the printed map. Possibly Waldseemüller intended for the blank spaces in the Americas to be later colored green, but the only surviving exemplar of his large world map is in black and white.

The chartmaker (or the illuminator) of *Carta del Cantino* posed the birds in three different ways and paid attention to the curvature of their beaks, the holding of their heads, and the coloring of their feathers. The parrots can be identified as the scarlet macaw (*Ara macao*), native to South America. The paintings, so true to life, suggest that the parrots on the chart had been studied directly. This is entirely possible, because by 1501 sailors had brought these striking birds to Lisbon from Brazil.[3]

Like the parrots, the trees on *Carta del Cantino* are drawn far larger than life, and they too are painted with care. The trees appear with the parrots in Brazil, and the chartmaker (or illuminator) painted trunks, branches, and leaves such that it is possible to differentiate three distinct types of trees in Brazil. As with the parrots, the trees are painted meticulously. Unlike the parrots, it is doubtful that either the chartmaker or illuminator had directly observed Brazilian trees, and therefore their presentation on the chart is stylized. Horizontal lines further suggest the lay of the land, and the use of blue and green conveys the thickness and depth of the forests. The visual imagery of parrots and trees invites the viewer to contemplate a distinctive place—with thick forests and beautiful birds—across the Atlantic Ocean.

Parrots repeat on Caverio's *Planisphère nautique* (1506) and again on Waldseemüller's *Universalis Cosmographia* (1507). Caverio painted four bright red birds sitting on top of trees in Brazil, while Waldseemüller placed a single parrot over empty space in the interior of Brazil. The parrot is simplified with each move (figure 5.1). Although Caverio used the same medium—paint on vellum—the parrots on *Planisphère nautique* (figure 5.1, *center*) are more stylized than on *Carta*

Fig. 5.1. Parrots Signify Brazil on Three Maps. Details from *Carta del Cantino*, 1502, Gallerie Estensi, Biblioteca Estense Universitaria, Modena, with permission from the Ministero per i Beni e le Attività Culturali e per il Turismo (*left*); Caverio, *Planisphère nautique*, 1506, Bibliothèque Nacionale de France, Paris (*center*); and Waldseemüller, *Universalis Cosmographia*, 1507, Geography and Map Division, Library of Congress, Washington, DC (*right*).

del Cantino (figure 5.1, *left*). Although Caverio retained the curved beaks and red feathers, the parrots are less detailed. His parrots are positioned individually, alongside of trees, whereas the close positioning of the three parrots on *Carta del Cantino* is more suggestive of a flock. The single parrot on Waldseemüller's world map is even more simplified (figure 5.1, *right*). Appearing in profile with neck markings and with dark wing, tail, and belly, Waldseemüller's parrot has a curved beak, an open eye, and feet with toes. Its neck markings do not appear in the painted forms on the parrots on either Caverio's *Planisphère nautique* or *Carta del Cantino*. As compared to the parrots on *Carta del Cantino*, Waldseemüller's parrot could be a generic bird of prey. Apparently sensing that the bird needed a label to identify it, Waldseemüller provided one: the parrot stands on a scroll with the label "Rubei psitaci" (red parrot).

Why parrots? In 1500, parrots *(Psittaciformes)* were exotic animals to Western Europeans, though Bruce Boehrer argues that Western Europeans tend to view them as magical, exotic, and wondrous when they are rare, but as pets or annoying pests when they are familiar. Although parrots once lived in Western Europe millions of years ago, no native parrots lived in Europe during recorded history. All parrots in Western Europe, therefore, had always arrived through trade. The parrots present in classical Greece and Rome came from India, but with the decline of this trade during the medieval era, parrots from India became extremely rare. This would change with the voyages of discovery, beginning with those to Africa, when parrots were once again brought to Europe. In 1500 parrots were still glamorous and astonishing to Europeans.[4]

The distinctive scarlet macaw was indeed unique. Parrots had been encountered in the Americas before 1500, and they had been brought back to Europe from islands in the Caribbean as early as 1493. The first Spanish chronicles contain references to the parrots seen during Columbus's first voyages to the islands and mainland of the Caribbean. Some of these accounts seem highly exaggerated; nevertheless, they do suggest that parrots were sighted by the Spanish as early as 1492. When Columbus reached the Caribbean, he received parrots as gifts. On the first voyage, Columbus saw a flock of parrots so large that it blotted out the sun, according to Bartolomé de Las Casas in his *Historia de las Indias*. In describing the second voyage of Columbus, Las Casas frequently notes the "infinite" and "different colors" of parrots, and he begins to link different birds with specific islands—for example, the green parrots of Dominica, or the parrots as large as roosters on Guadalupe. According to Las Casas and other chroniclers, parrots had practical value—most specifically that they were eaten throughout the Caribbean. Las Casas lists parrots as among the food available for consumption, just as he so designates fish or rabbits; while on the island of Cuba, Las Casas explains how parrots are captured; and he further claims that on one occasion more than ten thousand parrots were eaten in fifteen days.[5]

Parrots were among the first animals that crossed the Atlantic from the Americas to Europe. Las Casas cites parrots as among the first gifts Columbus presented to Ferdinand and Isabella. The chronicler Peter Martyr d'Anghiera wrote that Columbus brought forty parrots back to Spain following his first and second voyages. The chronicler Gonzalo Fernández de Oviedo y Valdés records that he was an eyewitness to Columbus's return in 1493 and was present when Columbus visited the king and queen in Barcelona, bringing with him many parrots, as well as the first native peoples.[6]

The Portuguese encountered parrots on the first documented landing on the coast of Brazil by Pedro Álvares Cabral, in April 1500. Only three documents survive to recount the ten days spent at Porto Seguro, and two of these describe parrots. Some parrots seemed familiar, though others were not. In Pedro Vaz de Caminha's long letter to the king, he recounts sighting nine or ten parrots—green and gray—when he was with a group of men who entered the forest to cut wood. An account written by an anonymous man on Cabral's voyage describes "parrots of many colors, among them are some as big as hens." The anonymous author notes that sailors exchanged rattles and paper for them.[7]

At least two of these large parrots, which seem to have been scarlet macaws, arrived in Lisbon, after having endured the long journey from Brazil to South Africa, from East Africa to India, and from India to Lisbon. Bartolomeo Marchionni, a wealthy merchant from Florence who was living in Lisbon, owned a share in the ship *Anunciada,* one of the smallest and fastest ships in Cabral's armada, and the first to return—on 23 June 1501. Marchionni closed his letter, written just a few days later—on 27 June 1501—writing: "They brought back two parrots of different colours which are an arm [*gomito*] and a half long which are more than an arm and a half of ours. They are marvelous things." In a letter to the Signoria of Venice, Giovanni Matteo Cretico discusses the arrival of the same ship and recounts the story of its voyage, including that on this voyage of more than fifteen thousand miles "above the Cape of Good Hope towards the west they have discovered a new land. They call it that of the parrots [*papaga*], because some are found there which are an arm and a half in length, of various colours. We saw two of these."[8]

So prevalent was the association of parrots with Brazil that Pietro Pasqualigo, the Venetian ambassador in Lisbon, called Brazil the "Land of Parrots" in his report to the Senate of Venice in October 1501. This name—"Land of Parrots"—was not the official name of Brazil; rather it appears to have been the name that circulated in Lisbon among mariners. Arriving during Holy Week in 1500, Cabral formally gave Brazil the name *Terra da Vera Cruz* (Land of the True Cross) at vespers on 22 April. The king of Portugal used the name Santa Cruz when announcing the discovery in a letter to the monarchs of Spain in 1501. The chartmaker of *Carta del Cantino* incorporated both names. When illustrating Brazil, he chose parrots, thereby conveying the knowledge of the men of the sea, whose eyewitness observations were deemed reliable, and capturing the exotic nature of the scarlet macaws that were known to live there. In the legend off the coast, however, the chartmaker supplied the formal name. The legend begins with the official name "A Vera Cruz" (The True Cross) followed by the sign of a cross—"✝."[9]

Although we cannot be certain that the chartmaker of *Carta del Cantino* was the first to place parrots over Brazil, he had to be among the very first. The first documented arrival of parrots arrived in Lisbon with the return of Cabral's first ships in 1501, while the return of the voyage that reconnoitered the long coastline of Brazil—which served as the basis for the depiction of the coastline of Brazil on *Carta del Cantino*—occurred in July 1502. Soon after that, by October 1502, Alberto Cantino had arrived with the world chart in Genoa. By picking the parrots for the illumination, the chartmaker followed a long-standing practice in both the chartmaking and *mappaemundi* traditions—that of selecting striking animals, known to be native, to represent specific locations. At the same time, the vivid representation of the scarlet macaws is due in part to the fact that he chose an animal that had made a transatlantic voyage and could be observed in Lisbon. The illumination made the distant land across the Atlantic seem closer, and it was a concrete example of the movement of peoples and animals already happening in the Atlantic.

Why does this visual imagery repeat and change slightly from chart to chart and from chart to map? The simple answer is copying, which is and was a common and usual part of mapmaking. Caverio's world chart has its own message and aesthetics, but it is clearly a close copy of *Carta del Cantino*. Waldseemüller, on the other hand, used Caverio only for certain parts of his world map—the Fourth Part of the World and West Africa. Geographic information also migrated, such as toponyms, as well as geographic markers. For example, the unique design of the Portuguese *padrões* on Caverio's *Planisphère nautique* was repeated on Waldseemüller's *Universalis Cosmographia*. On both the Caverio chart and the Waldseemüller printed map, the *padrões* (stone markers) are represented as crosses into which is set the Portuguese crest (figure 5.2, *center and right*). As described in chapter 1, the Portuguese erected *padrões* along the coast of the West and East Africa, and they often appeared on Portuguese charts, as their exact latitude had been calculated by mariners using the astrolabe. The *padrões* can be seen on *Carta del Cantino* as tall crosses, painted in beige tones (figure 5.2, *left*). Caverio modified the "look" of the *padrões* by inserting the Portuguese crest and painting it in red and blue. Waldseemüller's *padrões* resemble

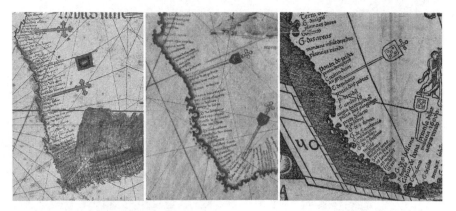

Fig. 5.2. The Portuguese *Padrões* on Three Maps. Details from *Carta del Cantino*, 1502, Gallerie Estensi, Biblioteca Estense Universitaria, Modena, with permission from the Ministero per i Beni e le Attività Culturali e per il Turismo (*left*); Caverio, *Planisphère nautique*, 1506, Bibliothèque Nacionale de France, Paris (*center*); and Waldseemüller, *Universalis Cosmographia*, 1507, Geography and Map Division, Library of Congress, Washington, DC (*right*).

Caverio's—although without the color—thus revealing that Caverio's modification was adopted by Waldseemüller.[10]

Because the parrot becomes substantially more simplified when it moves from manuscript chart (Caverio) to printed map (Waldseemüller), it raises the question of whether the medium was to blame. The chartmakers of *Carta del Cantino* and the *Planisphère nautique* worked with brush and paint—ideal for capturing the colors and textures of the parrots. Some simplification occurred as Caverio (or his illuminator) changed the position of the birds. Moreover, in Genoa in 1502 it was less likely that a scarlet macaw could be observed and sketched. On the *Planisphère nautique,* the beak, wings, and tail are fairly well reproduced, although the color is a different shade and the birds are smaller in size relative to the trees. When Waldseemüller chose to transfer the image of the parrot, could the medium of woodblock printing fully capture the essence of the parrot? Or was it necessary to simplify the image in the transition from manuscript chart to printed map? And did the artisans who—as seen in chapter 4—participated in the creation of Waldseemüller's printed map as designers and woodcutters have had access to other sources that could be used to design and cut the parrot? It seems even less likely that Waldseemüller or the artisans who worked on the printed map could have directly observed a scarlet macaw in either St. Dié or Strasbourg in 1507.

Printed images then in circulation could have provided examples copied by Waldseemüller's artisans. Dürer's *Adam and Eve,* dated 1504, includes a parrot

Fig. 5.3. Engraved and Woodcut Parrots. Details from Albrecht Dürer, *Adam and Eve*, 1504, Cleveland Museum of Art (*left*) and Martin Waldseemüller, *Universalis Cosmographia*, 1507, Geography and Map Division, Library of Congress, Washington, DC (*right*).

(figure 5.3, *left*). Dürer posed Adam and Eve in the Garden of Eden, before a grove of trees, with a parrot next to, and slightly behind, Adam's head. The parrot has a curved beak, long wings with feathers of different sizes, and a long tail. This parrot has been variously identified as a macaw, the same kind of parrot that appears on *Carta del Cantino*, and as a ring parakeet from India. The parrot on Waldseemüller's world map is strikingly similar (figure 5.3, *right*).[11]

Still, *Adam and Eve* was not a woodcut but an engraving. Both processes require printing with ink on paper, but the preparation of the base (the copper plate in the engraving and the woodblock in the woodcut) is not the same. In the copper plate medium, the plate was fairly soft and the artist, by varying the pressure, could use the burin to produce curves that swelled and tapered, as well as very thin lines. The result is a finer printed image, capable of preserving great detail, as compared to woodblock printing. Nevertheless, Dürer's woodcuts, as visible in the *Apocalypse* series, display careful attention to detail, and they exhibit a range of textures. Similarly, the many Habsburg eagles that appear on the *Arch of Honor*, circa 1515–19, are highly detailed and printed from woodblocks, as is the crest for the Behaim family, which features a detailed bird meticulously cut into a woodblock by Dürer or an assistant. By 1500 woodcutting was so well developed in Germany that the woodcut medium allowed for a range of textures and details. A skilled woodcutter could have reproduced a more detailed bird on Waldseemüller's map. The medium would have allowed it, but it seems not to have been important to do so.[12]

In its migration from a hand-painted world chart (*Carta del Cantino*) to the printed world map (Waldseemüller's *Universalis Cosmographia*), the original visual codes of parrots and trees slip. Details deemed to be important for *Carta del Cantino* are less prominent on Caverio's *Planisphère nautique* and even less so on the printed *Universalis Cosmographia* (figure 5.1, *right*). The simplification of the parrot illustrates that, as the image moved, the distinctiveness of the scarlet macaw was lost and replaced by a generic red parrot. Moreover, Waldseemüller made the parrot easier to "read" by supplying a label.

In the case of the *padrões*, the seemingly more complex images on the Caverio and Waldseemüller maps actually result from the combining of two separate images—flags and stone markers—both visible on *Carta del Cantino*. The adding of the Portuguese crest to the stone marker by Caverio simplifies the geographic knowledge, even as the resulting image becomes more complex. Whereas the chartmaker of *Carta del Cantino* distinguished between where the *padrões* were placed and where the Portuguese had established political claims, Caverio (and Waldseemüller after him) merged the flags with the stone crosses. Because it combines a religious symbol with a national flag, the result is a more powerful sign, but one that generalizes the nature of Portuguese influence in South Central Africa.

The migration and simplification of imagery from chart to chart and from chart to map can also be seen with trees. On the *Carta del Cantino*, the trees—of three distinctive types—painted over Brazil are shown in a receding landscape, suggestive of great depth (figure 5.4, *top*). On *Planisphère nautique*, Caverio greatly simplifies the trees and the landscape (figure 5.4, *center*). The landscape is even more abstract on Waldseemüller's *Universalis Cosmographia*. Waldseemüller's designer and woodcutter created mountains with closely cut sloped lines that convey both distance and height. Interspersed among these mountains are rounded shapes that seem to be trees (figure 5.4, *bottom*). These abstract trees shown in the Americas on Waldseemüller's world map are distinct from the trees in Asia and Europe. The European and Asian trees are highly stylized and generic, even as two distinctive trees exist—one a pine, the other a deciduous—and they both may have been created with a stamp. Only one unique tree, a palm tree, appears off the coast of West Africa. The migration of trees from Caverio to Waldseemüller is a simplification, but the abstractions might also have been intended to convey the lack of knowledge about the new lands. Or, perhaps, the abstract form of the trees was a choice made by artisans working with Waldseemüller. Abstract trees are not unique to the *Universalis Cosmographia,* as very similar trees appear in Dürer's woodcut of *St. Michael Fights the Dragon.*

The selection of trees as a sign for the Americas results from both eyewitness observations of trees and the recognition of their commercial potential. The first reports sent back from the Americas register not only the existence of vast forests but also the prospective use of American timber in shipbuilding. In Las Casas's

Fig. 5.4. The Abstraction of Trees. Details from *Carta del Cantino*, 1502, Gallerie Estensi, Biblioteca Estense Universitaria, Modena, with permission from the Ministero per i Beni e le Attività Culturali e per il Turismo (*top*); Caverio, *Planisphère nautique*, 1506, Bibliothèque Nacionale de France, Paris (*center*); and Martin Waldseemüller, *Universalis Cosmographia*, 1507, Geography and Map Division, Library of Congress, Washington, DC (*bottom*).

summary of Columbus's *Diario*, Columbus describes trees with care. He first notes the value of trees he identifies as mastic (*almaçiga*) and aloe. On 25 November, a Sunday, a new kind of tree was seen and described with great excitement. These were pines. The *Diario* states that the ships' boys "shouted that they saw pine groves," and when Columbus looked up and saw them, he was very impressed with their "height and straightness, like spindles, thick and thin." He immediately saw their value and thought the place could serve as a shipyard because of the "vast quantities of planking and masts for the greatest ships of Spain."[13]

Dyewood was another valuable trading commodity, quickly identified in the Atlantic islands and mainlands. *Brasile, brasil, brésil, braxile, verzino,* and similar terms with various other spellings all referred to a red dye extracted from trees. Originally, an East Indian tree produced the dye, but the name was extended to similar trees found in South America. The value of dyewood trees was noted by the observers of ships returning from the Americas to Spain and Portugal. Alberto Cantino wrote to Ercole d'Este that, on 5 October 1501, a Spanish ship returned to Cadiz loaded with sixty slaves, fifty marks of pearls, and three hundred *cantara* of both *braxilio* and *verzino.* Vespucci reported extensive stands of *verzino* in 1501. In his third letter to Lorenzo di Pierfrancesco de' Medici, Vespucci wrote that as the expedition sailed for eight hundred leagues along the coast "we found there endless brazilwood, of very fine quality, enough to load all the vessels on the seas today, and it costs nothing."[14]

In the far northern Atlantic, as depicted on *Carta del Cantino,* the chartmaker locates a place called "Terra Nova" (New Land). This land is so new that the chartmaker does not yet have access to the place names that correspond to its coastline. The chartmaker illuminates the land by painting a dense grove of trees. Terra Nova is Newfoundland, as letters written from Lisbon in October 1501, as well as the legend written directly on *Carta del Cantino,* make clear. It was discovered by the Corte Real brothers in 1500–1501. After the return of one of the brothers, Alberto Cantino wrote to Ercole d'Este that the land was covered by trees so tall and so thick that there would be more than enough tree trunks to supply all of the masts needed by the largest navy in the world. Similarly, Pietro Pasqualigo, writing to the Venetian Senate, described the king of Portugal as so happy with the news of the discovery of this land because its thick forests of pine and hardwood trees would supply timber for shipbuilding. On *Carta del Cantino* the legend next to Newfoundland hails the commercial potential of timber: "There are here many masts."[15]

The chartmaker of *Carta del Cantino* changed his mind about how to present Brazil sometime during the making of the world chart, and this change accentuates the importance of trees as a visual code. A patch on the world chart reveals that there was an earlier version of the coastline of Brazil that extended out farther to the east into the ocean. Presumably, when the exploratory expedition of 1501 returned to Lisbon in 1502, the chartmaker patched the vellum to correct the placement of the Cabo de São Jorge, which he had earlier located too far to the east. Although the chartmaker knew to correct Cabo de São Jorge, he seems not to have had access to the full list of toponyms that would have been generated by the exploratory expedition to Brazil. Moreover, the thickly painted vignette of parrots and trees already occupied the space where toponyms would normally have been written. Only two toponyms seem to have been planned: "Cabo de Sam Jorge" and "Porto Seguro." Both appear as carefully lettered in red ink. When the patch was made, some of the illumination was lost, and several toponyms were added in with a different, less careful hand on the ocean side: "Cabo de Sã Jorge" (Cabo de São Jorge), "San Miguel" (São Miguel), and "Rio de Sã Franc°" (Rio do São Francisco).

One more toponym appears, but on the land side, presumably because there was not room on the ocean side (where the chartmaker had placed the legend describing the discovery). This toponym is "Rio de Brasil" (River of Brazil), lettered in red and barely visible over the green paint. *Brasil* or *Brazil,* as noted earlier, referred to a red dye extracted from trees, and Vespucci had reported *verzino* (brazilwood) along the coast of Brazil. By the time the chartmaker patched *Carta del Cantino,* brazilwood had been identified as a likely export. With this toponym inserted, the trees painted over Brazil on *Carta del Cantino* gain a clear commercial value. Not only do the trees characterize the nature of the landscape, but they can also be read as stands of brazilwood.

The connection between imagery and commercial potential is also visible on *Carta del Cantino* with respect to Africa. Various images appear, such as parrots, a lion in Sierra Leone, an elaborate representation of the trading post at São Jorge da Mina, African villages, native Africans dancing, a scene of a gallows, and depictions of unique landscapes, such as the Atlas Mountains in the North and the Cape of Good Hope in the South. The African parrots are also native species: the green parrots painted over West Africa can be identified as the Senegal parrot (*Poicephalus senegalus*), found in Mauritania, Senegal, and Guinea, while the gray parrots with red tails painted in West Central Africa can be identified as the African gray parrot (*Psittacus erithacus*), found south of the Ivory Coast and into the Congo region.[16]

How the chartmaker of *Carta del Cantino* chose to illustrate Africa furthers our understanding of the meaning of the parrots in the Americas. The chartmaker already had a great deal of information about West Africa, as well as a growing corpus of knowledge for South Central Africa. The four Senegal parrots with predominantly green plumage opposite the place on the map labeled "Canaga" (Senegal) are not as large as the scarlet macaws in Brazil, nor are they shown in a painted landscape (figure 5.5, *left*). This is most certainly due to the fact that the chartmaker had taken up the space just inside the West African coastline to write in the toponyms. In 1502 the west coast of Africa was well known in Lisbon, and accordingly it is well labeled on the land side. As a result, less room remained for illuminations. The four African grays with red tails placed farther south, inland from the Bight of Benin, are larger than the Senegal parrots and are, like the scarlet macaws in Brazil, depicted alongside trees and

Fig. 5.5. Parrots Signify Africa on *Carta del Cantino*. Details from *Carta del Cantino*, 1502, Gallerie Estensi, Biblioteca Estense Universitaria, Modena with permission from the Ministero per i Beni e le Attività Culturali e per il Turismo.

fields (figure 5.5, *right*). Unlike the Brazilian landscape, the Benin landscape is inhabited. Round huts with thatched conical roofs suggest villages.

Near the African gray parrots, over Benin, a legend on *Carta del Cantino* reads: "Here is the King of Benin, this king is a Moor [i.e., Muslim], and the people are black, and they trade much with the ships that go from Portugal and with these islands, namely the Island of S. Tomé, etc., and from here they bring many slaves and gold and musk and other things and gray parrots and shells and pepper."[17] This legend implies that by 1502 parrots were commonly brought from Africa to Lisbon. In fact, when Cabral landed in Brazil in April of 1500, he had with him a gray parrot, presumably an African gray. Caminha writes that when Cabral received two native men on board his ship, he showed them the parrot. The men took the bird on their hands, showing that they were familiar with it, and gestured, Caminha writes, as if to say that parrots lived in their land, too. Thus, not only did the chartmaker appear to have selected the two species of parrots as emblematic of specific regions in Africa, but he (or the illuminator) likely had also observed the two African species in Lisbon. This allowed for the birds to be sketched and painted on the chart so that their distinctive markings would be visible. Each parrot had a different geographic range, which the chartmaker recognized and deemed important enough to place each in its respective habitat. At the same time, the chartmaker spelled out in the legend that they were also commodities. Forerunners to the scarlet macaws, the Senegalese parrots and the African grays had also crossed the Atlantic Ocean. As visual codes for West and West Central Africa, they, like the scarlet macaws, conveyed the interconnected nature of the Atlantic.[18]

Other legends similarly explain images and make clear that Atlantic trade is a major theme that underlies the illuminations selected by the chartmaker of *Carta del Cantino*. The most prominent vignette over Africa presents São Jorge da Mina, the fortified trading post begun by the Portuguese in 1482. The trading post is drawn like a city, and by presenting it in the same way as other prominent cities on the chart, such as Venice or Jerusalem, the chartmaker argues for its importance. The visual imagery is specific to this place: it shows the trading post with adjacent African villages with their distinctive huts. Native men and women—painted with black ink, some clothed, some not— dance before the fortified trading post of São Jorge da Mina. The legend reads: "From whence they bring to the most excellent prince Dom Manuel King of Portugal in each year twelve caravels [sailing ships] with gold; each caravel brings twenty-five thousand weights of gold, each weight being worth five hundred reis, and they further bring many slaves and pepper and other things of much profit."[19]

The legend emphasizes the significance and value of the trade at São Jorge da Mina, which is elaborately painted on the chart. And the legend reveals that slaves are shipped from the port. This opens up the possibility that the dancing figures on *Carta del Cantino* both symbolize the local people and define them as people who have been or will be forced into ships and transported across the Atlantic as slaves. The slave trade from West Africa to Western Europe was well established by the time the chartmaker selected dancing Africans as part of the visual code for São Jorge da Mina. Even the Portuguese Crown, while not the major trader in slaves, was deeply involved in the trade. The rate of the Portuguese slave trade in the late fifteenth and early sixteenth centuries has been estimated by economic historian Ivana Elbl at three thousand slaves per year. Thus the placement of Africans next to the lavish illumination of São Jorge da Mina, with the explanatory legend to confirm it, speaks to the present or future commodification of native Africans in the Atlantic slave trade. However, Caverio chose not to copy this imagery; instead, he reduced the illumination of São Jorge da Mina on *Planisphère nautique*, and Waldseemüller, in turn, indicated it only in the most rudimentary form. This suggests that this trading post was not as important to Caverio, and only minimally so to Waldseemüller.[20]

Both Caverio and Waldseemüller had many images to choose from when illustrating Africa, given that Africa—unlike the Americas—had been mapped repeatedly. As a result, the discernible pattern of visual movement for the visual iconography for the Americas—from *Carta del Cantino* to Caverio's *Planisphère nautique* to Waldseemüller's *Universalis Cosmographia*—does not occur for Africa. For example, the lion used to symbolize Sierra Leone on *Carta del Cantino* does reappear on the Caverio *Planisphère nautique* in a more stylized form, but it does not surface on Waldseemüller's *Universalis Cosmographia*. A giraffe appears on the Caverio, but not on either the *Cantino* or the *Universalis Cosmographia*. No elephant appears on *Carta del Cantino*, but an elephant appears on both Caverio's *Planisphère nautique* and Waldseemüller's *Universalis Cosmographia*.

On earlier maps of Africa, distinctive animals, including elephants, lions, and giraffes, had already been associated with specific places. In the *Atlas de Cartes Marines*, an elephant appears in northern Africa, and a hundred years later, on the anonymous Genoese planisphere of 1457, a large elephant similarly positioned near Nubia carries three rulers on his back in a howdah (litter), a giraffe signifies an inland kingdom, and a lion is associated with northern Africa. The rationale for the placement of imagery on this map derives, according to historian Angelo Cattaneo, from classical and medieval sources. Particularly influential was Isidore de Seville's descriptions of regions of Africa in his *Etymologies*. Seville writes that northern Africa "produces wild beasts, apes, dragons, and ostriches. Once it was also full of elephants, which now only India produces." Ethiopia, Seville writes, "also teems with a multitude of wild beasts and serpents. There, indeed, the rhinoceros and the giraffe are found."[21]

Using animals to symbolize specific places was well established in chartmaking and *mappaemundi* mapping traditions, and therefore animals were part of the traditional visual vocabulary available to chart and mapmakers when selecting visual codes. But whereas the chartmaker of *Carta del Cantino* preferred direct observation—as can be seen in the parrot illuminations for Brazil, Senegal, and Benin—Caverio and Waldseemüller turned to print sources when designing the elephant that each placed in South Africa.

Caverio's stationary elephant carries a howdah (in this case a pack), its body has the form of an ox, and the elephant's ear resembles that of a dog (figure 5.6, *left*). Waldseemüller's elephant has no howdah or pack and is shown walking alone with clearly delineated toes. The woodcutter outlined the elephant in black and filled the body in with repeated curving, black lines that suggest the mass of a huge animal. A large fan-shaped ear is carefully executed (figure 5.6, *right*). Comparing Caverio's elephant to Waldseemüller's shows that Waldseemüller or the artisans did not rely solely on the elephant on Caverio's *Planisphère nautique*. Instead, they made alterations in how the elephant was drawn and cut. These changes most probably derived from other printed images of elephants available to Waldseemüller or his artisans.

Fig. 5.6. Elephants. Detail from Caverio, *Planisphère nautique*, 1506, Bibliothèque Nacionale de France, Paris (*left*); Martin Schongauer, *An Elephant with Howdah*, ca. 1483, Cleveland Museum of Art (*center*); and detail from Waldseemüller, *Universalis Cosmographia*, 1507, Geography and Map Division, Library of Congress, Washington, DC (*right*).

A few printed images of elephants that circulated in Europe in the late fifteenth century probably provided both Caverio and Waldseemüller with sources on which to model their elephants. In Germany two engravings of elephants were created in the fifteenth century, possibly based on an elephant that was introduced there in 1483. Martin Schongauer engraved a print of an elephant sometime between 1484 and 1490 (figure 5.6, *center*). Schongauer's prints circulated widely, and they were collected, making it likely that this view of an elephant became part of the visual vocabulary available to many artists and mapmakers. Martin's brother, Ludwig Schongauer, also produced a print of an elephant. Ludwig's elephant is less faithful to the actual nature of the elephant—it stands still, has cloven hooves, and its body resembles an ox. It is led by a trainer wearing a turban and boots. Caverio's elephant seems to derive from the print by Ludwig Schongauer, for in his image a trainer, wearing a turban and boots, leads an ox-shaped elephant carrying a pack. Waldseemüller and his designers and woodcutters, on the other hand, produced an elephant closer to Martin Schongauer's engraving.[22]

In the migration of imagery from a hand-painted planisphere (*Carta del Cantino*) to the printed world map (Waldseemüller's *Universalis Cosmographia*), the parrot slipped, but in the migration of the elephant from Caverio's hand-painted *Planisphère nautique* to Waldseemüller's printed *Universalis Cosmographia*, the elephant became sharper. The truer-to-life elephant on Waldseemüller's *Universalis Cosmographia* did not necessarily result from Waldseemüller or the artisans having directly observed the beast but more likely from their familiarity with a print by an artist who had.[23] Because an elephant had already been placed in South Africa, and therefore was an established visual code associated with the region, print imagery could be consulted to render it on the map. Waldseemüller copied the idea of the elephant but not its form from Caverio's *Planisphère nautique*. The artisans (or Waldseemüller himself) took advantage of other materials to create the elephant, undoubtedly seeing printed images as valuable and reliable sources of information.

Visual codes are powerful on all maps, but how they emerged immediately after 1500 to depict and incorporate new places bordering the Atlantic Ocean is revealing. Eyewitness descriptions of flora, fauna, peoples, and landscapes could be transferred onto a chart made in a port city soon after a voyage returned. Using traditional paints and inks, chartmakers (or illuminators) quickly created visual analogies based on the impressions supplied by the returning sailors, captains, navigators, and pilots. This rapid visualization of the Americas can be seen in the two visual codes of parrots and trees. But these images did more than simply describe a place; they documented the circulation of peoples and animals through the Atlantic, and they pointed to the commercial potential of commodities found across the Atlantic. Legends that the chartmaker placed on *Carta del Cantino* illustrate how text and image worked together to further a chart's message. Subsequent chartmakers could copy visual codes or adapt them, especially when they had additional sources. Caverio, for example, copied the visual codes for the Americas from *Carta del Cantino*, but he had additional visual vocabulary available to him for Africa, as can be seen in his selection of the elephant. Waldseemüller adopted the elephant, but when he or the artisans who worked on the world map set out to design and cut the elephant, they found printed imagery that they deemed better than Caverio's painting. The design and execution of the elephant changed. Because printed images circulated more widely, they disseminated information more rapidly and more extensively than could a manuscript chart. In this way, printed maps—even when less accurate than a manuscript chart—became more influential in shaping knowledge about faraway places across the Atlantic.

In thinking about how the Americas were first represented in the Atlantic World, it is well to revisit Harley, who pointed out a major "silence" in the mapping of the Americas: the absence of the native peoples. On *Carta Universal, Carta del Cantino*, the *Planisphère nautique*, and the *Universalis Cosmographia*, no indigenous peoples appear. For Harley, this oversight is part of the colonial discourse, which made Native Americans "invisible in their own land." The cartographer's use of empty lands, coupled with the absence of native peoples, Harley argues, promoted the arguments in favor of colonies.[24]

Is the absence of native peoples in the early visual codes an argument for colonization, as Harley suggests? One can argue that it was not customary to depict native peoples in the genre of portolan chartmaking. None of the surviving

Portuguese charts made circa 1500 shows native peoples, with the exception of *Carta del Cantino*, which shows Africans near São Jorge da Mina. Similarly, in the Ptolemaic tradition native peoples are also not visually depicted on maps. Yet the *Universalis Cosmographia* incudes native peoples in South Africa. The presence of native peoples is a new visual code emerging on world charts and world maps after 1500. There would soon be an image that would become the dominant trope for the native peoples of South America. Although there is no sign of this image on *Carta del Cantino*, or on Caverio's *Planisphère nautique*, or even on Waldseemüller's *Universalis Cosmographia*, it would soon be ubiquitous. It was the cannibal.

CHAPTER SIX

The Cannibal Scene

Parrots, trees, and the color green are the earliest known visual codes used by chartmakers to represent the newly encountered lands in the Western Atlantic, but a far more damning image soon became ubiquitous: scenes of cannibalism. "Since the age of Columbus," J. B. Harley has stated, "maps have helped to create some of the most pervasive stereotypes of our world," and the association between cannibalism and Brazil confirms this observation. Through an extensive study of more than two thousand maps, Surekha Davies has found that cannibalism "was the most frequent motif for Brazil on maps well into the seventeenth century." As the Atlantic World opened, descriptions of newly encountered peoples began to circulate—orally, in printed texts, and as images. Sensationalized and unreliable reports about cannibalism found their way onto charts and maps after 1500 and reinforced arguments for what would become central features of the Atlantic World: colonialism and slavery.[1]

The first scene of cannibalism in the Western Atlantic appears on an anonymous world chart, or *Weltkarte*, known as *Kunstmann II* and the *Four Finger World Chart*, made between 1502 and 1506 (plate 12). The scene is prominent and violent. It consists of two men, both naked, with one turning the other's impaled body over a long fire (figure 6.1). No one looking at the map could miss it, especially when the red pigments used for the rising flames beneath the body—today faded—would have been vivid and bright. Next to the image, a place name is inscribed on the coast: *Cabo de Stancta* + (Cape of Santa Cruz), which can only be Brazil. Its chartmaker remains anonymous, but as it is considered to be a Spanish (or possibly an Italian) chart, Vespucci is one of several conceivable authors. It was created for—or subsequently owned by—Cardinal Bernardino Lopez de Carvajal, whose crest appears as part of the Garden of Eden, which is a prominent illumination placed over South Africa. Carvajal, formerly bishop of Cartagena (in Spain) was made cardinal in 1493 and served Ferdinand and Isabella in Rome. A learned man cognizant of the importance of Columbus's discoveries for Spain, he commissioned—or received as a gift—an illustrated world chart with a prominent image of cannibalism placed in the Americas.[2]

The image of cannibalism on *Weltkarte* is accompanied by a legend. Visual codes on maps and charts are not necessarily accompanied by legends and labels,

Fig. 6.1. Cannibalism in Brazil. Detail from *Portulan (Weltkarte)*, 1502–6. Bayerische Staatsbibliothek, Munich, Cod.icon. 133.

which are part of what Denis Wood calls the linguistic code of a map. This code consists of explanatory texts, names, and the map's legend. According to Wood, cartographers use the linguistic code to label, to clarify, to explain, and to high-light that which might not be immediately apparent to the map viewer. On *Weltkarte,* a legend placed in the same general area of the world chart states: "This land that was discovered is called the land of the Holy Cross because it was dis-covered on the day of the Holy Cross. And on it is a very great supply of brazil-wood. Also found there is cassia [wild cinnamon] as large as a man's arm. Parrot birds large as falcons. And red men almost white having no law eat each other."[3] At about the same time as the *Weltkarte*, a printed image that depicted can-nibalism among peoples living on lands across an ocean appeared in Germany. It was a broadside printed between 1503 and 1505. It survives in two copies: one at the Bayerische Staatsbibliothek in Munich, which dates it to 1503, and the other at the New York Public Library, which dates it to 1505. The print (34×25 cm) is set up horizontally (landscape view), and both extant copies are tinted (figure 6.2). The woodcut has been traced to Augsburg, where some twenty-four printers were active in 1500, and one of them—Johann Froschauer—is thought to have printed and sold it. We shall refer to it as the *Augsburg Broadside*.[4]

The *Augsburg Broadside* shows a scene—presented as if the artist were the eyewitness observer—dramatizing the lifeways of peoples living across an ocean. Cannibalism is a fact of daily life: a head, leg, and arm hang from a roof beam suspended over an open fire; a leg rests on a table, and a man gnaws on an arm. Moreover, cannibalism is so normalized and insinuated into everyday life that a woman nurses a baby and tends to older children at the same time. The *Augsburg Broadside* has four lines of text printed in German in black gothic type underneath the scene. The text begins: "This image shows us the people and the islands discovered by the Christian king of Portugal or his subjects."[5]

Broadsides (or broadsheets) were single sheets, printed from woodblocks, with an image and a few lines of text. A common broadside was the devotional print, often accompanied by text. Contemplating devotional images was a spiritual act—the image allowed viewers to imagine themselves within the experience depicted.

Fig. 6.2. Johann Froschauer, *Woodcut of South American Indians [Augsburg Broadside]*, 1505. The New York Public Library.

The image could be "read" whether one was literate or not, and the accompanying text, usually presented in the vernacular, could be recited to small groups of people. Another common theme for broadsides in the sixteenth century was the monster, understood not as a being living in a distant land but as an abnormality of nature, such as unusual births. According to Katherine Park and Lorraine Daston, "monster broadsides began with a provocative title, a schematic woodcut of the child or animal involved, and a brief description of the circumstances of its birth." Such broadsides opened up reflection on a variety of religious beliefs, such as God's wrath or the coming Apocalypse.[6]

The use of text in combination with imagery was characteristic of broadsides. As an example, Dürer's famous woodcut of the rhinoceros (1515) not only resembles in size the *Augsburg Broadside* but also consists of an image and text. Dürer's broadsheet (which measures 23.5 × 30 cm) includes above the animal five lines of gothic type. Just as in the case of the *Augsburg Broadside*, Dürer's text explains in German what the image shows. "With this broadsheet," writes Daniel Zolli, "Dürer spread the news of the newly imported pachyderm to a public with a taste for the sensational and exotic." Similarly, the *Augsburg Broadside* carried a striking message about previously unknown peoples living in newly discovered lands.[7]

The designer and woodcutter of the *Augsburg Broadside* were certainly artisans who lived and worked in the midst of other designers, printers, woodcutters, and typesetters, making it unlikely that the ideas and techniques used in this print emerged in isolation. Those who worked on the broadside would have belonged to local artisan guilds and been familiar with the flourishing market for books and prints. Yet the broadside depicted a scene that was unlikely to have been seen by any of them at first hand. Where the scene takes place is not explicitly stated, but the text clearly points to a place discovered by the Portuguese, and the viewer can see that one of the ships coasting offshore has crosses embroidered on its mainsail—recognizable in Atlantic ports as insignia of the Portuguese Order of Christ. Because the native peoples are described in the text and shown in the woodcut image as having long hair and brown skin, the ships may be presumed not to be off the coast of Sub-Saharan Africa, which makes the unidentified place somewhere in the western Atlantic or in the Indian Ocean. Suggestive of Brazil are the feathers and featherwork—skirts, headdresses, collars, and bracelets—that adorn every body in the image (except the nursing baby). A simply drawn and cut bird that might be a parrot appears in the center left of the image—between the mother and her child.[8]

Unlike the images of parrots and trees discussed in chapter 5, the manuscript and print images of cannibalism on the *Weltkarte* and the *Augsburg Broadside* do not resemble each other. The texts that accompany each image are also different. The text for the *Augsburg Broadside* emphasizes a description of the peoples beginning with their bodies, progressing to their habits, and ending with their mode of government. The legend on *Weltkarte*, on the other hand, presents the name of the place, its trading goods, and a distinctive animal (the parrot), and it ends with a brief reference to its peoples, mentioning their skin color, lack of laws, and cannibalism.

Despite their differences, these oldest surviving images of cannibalism in the Americas—the one drawn, cut, and printed in Germany; the other painted by hand on a world chart most likely in Spain—present cannibalism as a fact. The chartmaker powerfully conveyed this message by situating the scene in a specific geographic place on a world chart. This encouraged the map viewer to read the image of cannibalism as true, and the elements of charts—such as compass roses, rhumb lines, and toponyms—reinforce the interpretation that "this is what is to be found here." Although the broadside is not a map, it too proclaimed a truth. Several elements that are evocative of a map—the coastline, the water signified by wavy lines, the ships offshore—reinforce to the viewer that the broadside presents known facts. The detailed text that explained the image also had the effect of buttressing the truth depicted in the image.

The seemingly simultaneous emergence of cannibalism in manuscript and print poses provocative questions. Who was responsible for such negative imagery? Was it true? Why did images of cannibalism not appear first on manuscript charts? From where did the chartmaker and artisans in Germany obtain their visual and textual information? There are no definitive answers to these questions; however, in exploring them the power of texts in shaping the imagery used to represent the Americas after 1500 becomes clear.

Let us begin with the origin of the word cannibal. The word dates to 1492. Previously, as Peter Hulme explains, Western Europeans used a Greek word for humans eating other humans, which comes down to us today as "anthropophagy." As Hulme further elucidates, the two words—anthropophagy and cannibalism—both once referred to specific peoples who practiced it. The Greek word is "made up of two pre-existing words ('eaters/of human beings') and was bestowed by the Greeks on a nation presumed to live beyond the Black Sea." The more recent term is "a non-European name used to refer to an existing people—a group of Caribs in the Antilles."[9]

Anthropophagy was not unfamiliar to Western Europeans, for it had long appeared in texts, in visual imagery, and even on maps. It was often associated with monsters. In a literary tradition beginning at least as early as Homer, the hero Odysseus faces an anthropophagous monster—the Cyclops—who must be defeated. In Greek and Roman histories, anthropophagous peoples are encountered far away. Herodotus describes such peoples inhabiting a distant region beyond the Scythian farmers, although Pliny the Elder claims the Scythians themselves were anthropophagous. In the Middle Ages, tales of epic confrontations between knights and monsters became popular, while texts on monsters, such as the *Liber monstrorum* and the *Tractatus monstrorum*, continued to locate monsters in distant places. Among these monsters were barbarians and enemies who were believed to be anthropophagous.[10]

As more Western Europeans traveled to the East in the late Middle Ages, their tales extolled exotic wonders, even when they acknowledged anthropophagy. When these travel accounts were illustrated, artists often created monsters. For example, in the *Livre des merveilleus du monde* (1413), which includes a version of Marco Polo's travels, dog-headed monsters appear in his description of the inhabitants of the islands of Andaman in the Bay of Bengal. Marco Polo described the inhabitants as ugly and animal-like; he also described them as cruel people who ate human flesh raw. In the *Livre*, an artist created dog-headed monsters to accompany the text. According to literary scholar Michael Palencia-Roth, "Working from this description, the artist . . . found that he could best illustrate Marco Polo's simile as a metaphor; he then reified the metaphor. . . . That is, the people's heads are not merely *like* dogs' heads; they *are* dogs' heads." Marco Polo's popular text circulated in many manuscript copies throughout Western Europe; Columbus, for example, owned a copy. Medieval Spanish texts, such as the *Libro del conosçimiento,* suggest that in Iberia, as elsewhere in Western Europe, stories of monstrous peoples in distant places were well known.[11]

Monsters were a common trope on *mappaemundi*, appearing on the margins of the known world—such as the monstrous human races in East Africa on the Hereford *Mappa Mundi*. Mapmakers located peoples believed to be wild, fearsome, and eaters of human flesh in places far removed from Western Europe. For example, in northern Asia, mapmakers typically placed Gog and Magog—the dreaded peoples associated with the Apocalypse. In texts inserted onto their maps, mapmakers often described Gog and Magog as anthropophagous peoples and located them on a peninsula, or behind mountains, or held back by a wall.[12]

By the second half of the fifteenth century, some mapmakers began to question whether monsters truly inhabited the peripheries of the known world, or whether Gog and Magog referred to actual peoples. Legends on Fra Mauro's *mappamundi* suggest that he doubted the presence of monsters in Asia and Africa, as well as the story that Gog and Magog lived behind a wall erected by Alexander the Great. Instead, Fra Mauro implied that Gog and Magog should be understood allegorically, and he trusted instead the reports of Venetian merchants who had visited the Caspian Mountains. On the *mappamundi* in the

Nuremberg Chronicle—the richly illustrated encyclopedic summary of the world's knowledge printed by Hartmann Schedel right at the end of the fifteenth century—Gog and Magog are not mapped, nor are the monsters discussed in the text. Seven monsters serve as a kind of exterior border for the book's *mappamundi*—they are not integrated into the map itself. As art historian Stephanie Leitch reflects, the monsters that were once thought to inhabit the margins of the earth have been divorced from their geographic spaces and serve instead as decoration.[13]

With Columbus's return in 1493 came the first accounts of anthropophagous peoples in the Americas. The presence of monsters and anthropophagous peoples on earlier *mappaemundi*, as well as the belief that monsters lived in peripheral regions, has led some scholars to suggest that Columbus expected to find monsters and anthropophagous peoples when he sailed west in 1492. And scholars emphasize that, as there was no shared language, Columbus's descriptions are based on his interpretation of encounters that he poorly understood. Columbus's letter to Luis de Santángel—published after his return, in Barcelona in April 1493—associated anthropophagy with peoples encountered on the first voyage. In the letter, which circulated widely in Spain and beyond in the last decade of the fifteenth century, Columbus makes two statements about monsters, one of which alludes to anthropophagy. He writes "In these islands up to now, I have not found misshapen men as some people might have expected." Instead, Columbus characterizes the people as "of very pleasant bearing" with "straight flowing hair." The second comment dismisses the possibility of monsters but invites the contemplation of anthropophagy. Columbus writes, "of monsters, I haven't had the slightest report, except for one island that is here, the second island in the gateway to the Indian Islands that is settled by a people who are held among all the islands to be very fearsome; these people eat human flesh."[14]

These brief statements have a deeper backstory that appears in Columbus's log. Known as the *Diario de a bordo*, Columbus's log survives only in summary form, but it is deeply revealing about how anthropophagy became associated with the native peoples of the Caribbean. Columbus gave his log to the Ferdinand and Isabella, and Bartolomé de las Casas summarized the *Diario*—from a copy of the original—many years later. As Las Casas clearly states at the outset, the *Diario* is a summary and not an exact copy: "This is the first voyage and the courses and way that the Admiral Don Cristóbal Colón took when he discovered the Indies, summarized except for the prologue that he composed for the king and queen, which is given in full."[15] In the *Diario*, Columbus hears and records names of groups of native peoples—*caniba, canima,* and *canibales*—from which emerged the term *canibal*.

The *Diario* reports a number of entries where Columbus reflects on monsters and anthropophagy. Along the northern coast of eastern Cuba, Columbus learned from his indigenous guides about anthropophagous peoples called Canibales living on the island they called Bohío. In this entry for 23 November, according to Las Casas, Columbus understood his guides to be saying that these peoples had one eye in their foreheads: "The Admiral says that well he believes there is something in what they say." On 26 November, the information is repeated in the log, with the addition that the peoples who lived on the island of Bohío were called Caniba or Canima, and they were greatly feared because they carried away captives who never returned. The guides further described them as having "but one eye and the face of a dog." Las Casas notes that "the Admiral thought they were lying" and believed that the peoples they feared were "under the rule of the Grand Khan [Great Emperor of China]."[16]

In the entry for 11 December 1492, Las Casas records that Columbus understood from his guides that Bohío was much larger than Cuba, not surrounded by water, and possibly a mainland and that they called it Caritaba. Las Casas writes, paraphrasing Columbus, that "perhaps they are right for they may be oppressed by cunning people, because the people of all these islands live in great fear of those from Caniba." Las Casas then records Columbus's view that the Caniba were the "people of the Grand Khan" who lived nearby and came in canoes to capture the islanders. Because those captured never returned, "the other islanders think that they have been eaten." Yet another reference links Caniba or Canibales to anthropophagy; it appears in the log for 17 December, when some of Columbus's men are presented with arrows and told they are those of the Caniba or Canibales. Las Casas, summarizing Columbus, writes: "Two men showed the Spaniards that some pieces of flesh were missing from their bodies, and they gave the Spaniards to understand that the cannibals had eaten them by mouthfuls."[17]

The connection between Caniba, Carib, and anthropophagy comes during the last weeks of the first voyage. In the log for 26 December, after a dinner and conversation with a local indigenous chief, one of Columbus's archers fired a Turkish bow to impress the chief. The log then refers to an earlier conversation when "the men of Caniba, whom they call Caribs," carried bows and arrows and came to capture them. Las Casas records that "the Admiral [Columbus] told him [the chief] by signs that the sovereigns of Castile would order the Caribs destroyed."[18] Several days later, on 13 January, some of Columbus's men were on shore gathering yams when they met men with bows and arrows, one of whom was persuaded to board Columbus's caravel. Las Casas writes, "The Admiral [Columbus] says that he was quite ugly in appearance, more so than others that he had seen. He had his face all stained with charcoal. . . . He wore all his hair very long. . . . And he also was as naked as the others. The Admiral judged that he must be from the Caribs who eat men."[19] In the same entry, Las Casas records, "The Admiral [Columbus] says further that on the islands passed they were greatly fearful of Carib and in some they called it Caniba, but in Hispaniola, Carib; and they must be a daring people since they travel through all these islands and eat the people they can capture."[20]

Columbus links Carib and anthropophagy again, in the same entry, when describing the return of this same man—the one with his face "stained by charcoal"—who had been received on Columbus's ship. On shore, there were fifty-five men similarly dressed whom the Spaniards attacked. When Columbus heard of this, as recorded by Las Casas, "he was pleased because now the Indians would fear the Christians since without doubt the people there, he says, are evildoers and he believed they were people from Carib and that they would eat men."[21]

Carib becomes a place where its peoples, also known as Caribs, are anthropophagous peoples, on the very last days Columbus spent in the Caribbean. On 15 and 16 January 1493, according to Las Casas's summary, Columbus wrote that the island of Carib had much copper, but that it would be difficult to get because the people "eat human flesh." Columbus resolved to visit the island "where the people are of whom those of all those islands and lands have so much fear" because, Las Casas writes, "with their numberless [innumerable] canoes they travel through all those seas and . . . eat the men that they can catch." Columbus set his course for this island, according to what he understood from his guides, but when the island appeared to be more distant than expected, he resolved to take advantage of favorable winds and sail instead for Spain.[22]

On the journey back, Columbus drafted several letters, all of them lost except for the letter to Luis de Santángel. In this letter, Columbus summarizes his voyage, and the detail visible in the *Diario*—especially with respect to the names of the groups of peoples he met—is left out. His brief asides about not having seen monsters or that a group that he never met practiced anthropophagy contrast with a much more positive tone that he adopts in the letter to describe the peoples he encountered. Nevertheless, the printed Santángel letter did spread Columbus's assessment that anthropophagy existed in the Indies.

From the *Diario*, historians can glimpse some of the information that might have been narrated in the oral accounts of mariners. Their stories about the islands visited on Columbus's first voyage reached Lisbon—where Columbus initially landed—and subsequently Spain. Such stories did not circulate in print, but they became part of the informal knowledge shared in maritime communities. This oral knowledge would become familiar to chartmakers who lived in ports, and indeed some, like La Cosa, were on that first voyage. The influence of oral knowledge can be seen in the letter written to the *cabildo* (town council) of Seville by Diego Alvarez Chanca, who was on Columbus's second voyage and who reported anthropophagy immediately, suggesting that he anticipated finding it. After a quick, twenty-day Atlantic crossing from the Canary Islands, Columbus's large second fleet of seventeen ships reached islands south of where he had landed on the first voyage. Chanca describes these islands—thought today to be Dominica and Guadeloupe—as beautiful, green, and mountainous. Columbus ordered a boat to land on Guadeloupe, and its captain returned to report, according to Chanca, that he found houses with cotton, food, parrots, and "four or five human arm and leg bones." Chanca writes that this supported the view that they had arrived at the "Caribe" islands, "which are inhabited by peoples who eat human flesh," thereby confirming his familiarity with Columbus's view that anthropophagous peoples lived on the islands. Chanca subsequently describes Carib peoples as "those who eat human flesh" and as living on the islands that had been pointed out to Columbus on the first voyage.[23]

A second account of anthropophagy on the second voyage appears in a report based on a letter by Guglielmo (or Guillermo) Como, who may not have been on the voyage. Niccolò Scillacio, who wrote the report in 1494, stated in the introduction that he had received correspondence from Como and that he had translated it into Latin. In Scillacio's report, anthropophagy is described immediately on arrival in the Spanish Indies. In the English translation by John Mulligan, the text states: "These islands are under the rule of the Caribs. This barbarous and indomitable race feeds on human flesh. I might with propriety call them anthropophagi." In Scillacio's original Latin, however, the islands are called "Insulae cannaballis" (Islands of the Cannibals), the indigenous group is named "cannibali, and he refers to them as "anthropophagos." Scillacio's printed report thus makes a direct association between cannibal and anthropophagy.[24]

Given these textual sources, one might expect that cannibalism would appear on early charts, whether in legends, in images, or at least names. The toponym "canibal" does appear on the earliest surviving charts as a name for several different islands—labeled variously as "Ilha de los canibales," "Canibales," "Islas de Canibales," or "Canibalos." But no text or image conveys that the toponym "canibal" means anthropophagy. Because La Cosa was on Columbus's first and second voyages, he was familiar with the encounters that took place on the first, as well as with the oral knowledge shared by sailors. On the charts he must have made (but that no longer exist), he would have outlined islands and suspected mainlands, and he would have selected and entered toponyms. On the world chart that does survive, according to Silió, La Cosa included many indigenous toponyms for the Caribbean, and he did not always use the names first given by Columbus. On *Carta Universal*, islands, labeled "Islas de Canibales," appear at the entrance to the Caribbean.[25]

Two years later, on *Carta del Cantino*, there is an island labeled "Ilha de los canibales" (*Island of the Cannibals*) located just North of the South American mainland, near the location of present-day Aruba. This island has moved from where it appeared on La Cosa's chart. The *Portolan Chart (King-Hamy)* just presents an outline of the coastline of the Caribbean and South America (plate 5); there are no toponyms supplied. Caverio's *Planisphère nautique* repeats the same toponym as on *Carta del Cantino*—"Y de los canibales"—and locates it in the same place, as does Waldseemüller, who uses the toponym "Canibales," on his printed world map: *Universalis Cosmographia*.

Although the toponym "canibales/Canibalos" appears on the earliest surviving world charts and maps, nowhere on them is there a text that explicitly connects anthropophagy with these names or places. Rather, the legends written on the charts and maps proclaim rights of discovery, often citing the leader of the expedition, the crown that financed the expedition, and the potential trading goods to be had. The linguistic code on La Cosa's *Carta Universal* is minimal, and there is no mention of anthropophagy anywhere in the Western Atlantic. A short legend placed off the coast of North America states "sea discovered by the English," which refers to the voyage by John Cabot. No legends appear in the Caribbean, but a legend opposite Cabo de São Roque in South America states: "This cape was discovered in the year 1499 by Spain and the discoverer was Vicentinas," which refers to the expedition led by Vicente Yánez Pinzón in 1500. The linguistic code is more extensive on *Carta del Cantino*, which opens up the possibility that had the chartmaker wished to include a statement on anthropophagy, it would have been possible to do so. However, the text applied near the toponym for the "Ilha de los canibales" (Island of the Cannibals) begins with "the Antilles of the King of Castile" in a large hand-lettered script and then continues in smaller letters to say: "discovered by Columbus, who is admiral of them, which said islands were discovered by order of the most high and powerful prince King Dom Ferdinando King of Castile." As noted earlier, on *Carta del Cantino* the island labeled as "Ilha de los canibales" has moved from where it appeared on La Cosa's *Carta Universal*, and in its new location, off the coast of northern South America, a nearby island is labeled "ylha dos gigantes" (island of giants), possibly suggestive of monsters, while along the coast there is a toponym suggestive of wild peoples—"Costa de gente brava" (Coast of the wild people). Yet, there is no explanatory text that describes monsters or anthropophagy. Caverio's *Planisphère nautique* repeats the same toponyms found on *Carta del Cantino*, the only difference is a slight variation in spelling: "Y de los canibales," "Insula de gigantes," "Costa de gente brava." The locations of these toponyms are the same as on *Carta del Cantino*, and the legend proclaiming Columbus's discovery and the role of the king of Castile is virtually identical. Similarly, on *Universalis Cosmographia* Waldseemüller presents the same toponyms in the same places— "Canibales," "insula de gigantibus," "Costa de gente bravo." Waldseemüller's legend is even shorter: "These islands were discovered by Columbus, Genoese admiral sent by the king of Castile."[26]

A striking feature of *Weltkarte* is not only the scene of cannibalism in South America but an accompanying legend. The chartmaker inserted the legend into a scroll-like frame and placed it in the same general area. The legend confirms what the image shows: that the people practice anthropophagy. Previous charts and maps also placed legends in South America, but none had described anthropophagy. Rather, they recorded the discovery of the mainland and its perceived nature. For example, the legend for Brazil on *Carta del Cantino*:

> The True Cross + called by this name, which was found by Pedro Álvares Cabral, a nobleman of the house of the King of Portugal, and he discovered it when he went as captain-major of fourteen ships which the said King was sending to Calicut and going this way he met with this land here, which is believed to be a continent, in which there are many people who go about naked as their mothers delivered them; they are more white than brown and have very lanky hair. This said land was discovered in the year [1]500.[27]

The legend for Brazil on Caverio's world chart is virtually identical to that on *Carta del Cantino*, while Waldseemüller added a little more information on *Universalis Cosmographia*. Waldseemüller notes the discovery, suggests that the land is an island, and describes people who wear no clothes, but he makes no mention of anthropophagy:

> To the captain of fourteen ships that the king of Portugal sent to Calicut this place first appeared: which was thought to be mainland, when in truth it is, along with the part discovered earlier, an island surrounded by water, of incredible but not yet entirely known magnitude. In there the people of the male and even the female sex were accustomed to go [around] no differently than their mother birthed them. And they are here indeed a little more white than those whom they discovered on an earlier voyage, a deed accomplished by the command of the king of Castile.[28]

One might ask if the chartmakers before *Weltkarte* did not believe in monstrous peoples living in distant places and whether this accounts for the absence of images of anthropophagy and monsters in the Western Atlantic. However, the surviving charts and maps before *Weltkarte* do have images and labels that refer to monsters and anthropophagous peoples, but not in the Americas. On *Carta Universal*, La Cosa (or others working on the chart) wrote "R. Got" and "R. Magot" (King Gog/King Magog) in the far Northeast of Asia. Both labels have an accompanying illumination that associates the kings with familiar monsters—Gog as king of Cynocephali (i.e., of the dog-headed peoples) and Magog as king of Blemmyae (i.e., of peoples with no heads). On *Carta del Cantino*, Andaman Island, in the Bay of Bengal, has a legend that notes that the peoples living there are anthropophagous. There are monsters on Waldseemüller's *Universalis Cosmographia*—they stand partially hidden amid peoples in South Africa. The group, composed entirely of men, is drawn in contrapposto, with depth to their bodies. One man stands with a long staff or club and another holds a bow. The men are humans and not monsters, and the group seemingly depicts the native peoples of the Cape of Good Hope who were encountered by Portuguese expeditions beginning with Bartolomeu Dias. Among these men are two monstrous creatures at the back. (figure 6.3).[29]

If chart and mapmakers believed that barbarous peoples lived in the world's peripheries, why not project them into Western Atlantic? Moreover, because reports following Columbus's first and second voyages had described cannibalism, why did chartmakers not create anthropophagous images? That anthropophagy was not described in legends or visualized on early maps of the Western Atlantic suggests that early chartmakers and cosmographers did not see it as essential. Possibly they suppressed the knowledge—either because they found it unreliable or because it did not fit into the goals of their maps and charts. In mapping the Caribbean, chart- and mapmakers used as a toponym a name that originated

Fig. 6.3. Two Monsters among Men in South Africa. Detail from Martin Waldseemüller, *Universalis Cosmographia*, 1507, Geography and Map Division, Library of Congress, Washington, DC.

with Columbus's first voyage, but they did not enter explanatory texts that included descriptions of anthropophagous peoples. In mapping Brazil, the chartmaker of *Carta del Cantino* used information that returned with Cabral's ships. In the three surviving accounts of Cabral's landing in 1500, anthropophagy is not described, and presumably the oral accounts of mariners did not relate it either. Beginning with *Carta del Cantino* and continuing on to Caverio's *Planisphère nautique* and Waldseemüller's *Universalis Cosmographia*, the information conveyed on the maps about Brazil made no mention of cannibalism.

When did the earliest known printed world maps include mention of anthropophagy in the Americas? Based on surviving printed maps, a legend appeared first, and an image later. The first known textual legend can be seen in an edition of Ptolemy's *Geography,* published in Rome in 1508. The mapmaker, Johannes Ruysch, included a legend that described anthropophagy, but he did not insert a visual image of it.[30] Opposite Brazil on Ruysch's world map is a long legend that does not emphasize the date of discovery or the right of possession. Instead it consists of a description of the native peoples, animals, trading goods, and cannibalism:

> Land of the Holy Cross / This region, which many believe to be another large continent of the world, is inhabited in scattered fashion. Females and males walk around either completely naked or ornamented with woven roots and birds' feathers of various hue. Many live together with no religion or king. They fight wars between themselves without stopping, and they devour the human flesh of their captives. They enjoy weather so mild that they live over 150 years. Rarely are they sick and then they are cured with only roots of herbs. Lions breed here, and serpents and other foul beasts. Forests, mountains and rivers are here as well as a great abundance of pearls and gold. The Lusitanians [Portuguese] carry away from here brazilwood, also known as verzini, and wild cinnamon. / Or the New World.[31]

Every element but one in Ruysch's legend appears in Vespucci's first published letter—the *Mundus Novus*. Published in Latin in 1502 or 1503, the *Mundus Novus* letter was widely reprinted throughout Europe and translated into many languages. It presented a detailed description of anthropophagy and associated it with South America. Moreover, the letter described exactly where this *mundus novus,* or new world, was. The letter explains that it is reached by sailing from Cape Verde into the Atlantic Ocean, seven hundred leagues west but toward the Antarctic. Vespucci situates this new world eight degrees south of the equator and extending well south, beyond the Tropic of Capricorn, to 17.5 degrees from the Antarctic Circle. This precise geographic location of place is reinforced with Vespucci's claims of eyewitness authority: "We sailed along the shore, approximately six hundred leagues, and we often landed and conversed with the inhabitants of those regions, and were warmly received by them, and sometimes stayed with them fifteen or twenty days at a time."[32]

Ruysch frames his textual legend between two larger titles, "Terra Sancte Crucis" (Land of the Holy Cross) written above and "Sive Mundus Novus" (or New World) written below. Ruysch also describes the region as "another large continent." Vespucci coined the phrase *mundus novus*, and he claims to have discovered a new continent in the first few lines of the *Mundus Novus*. Ruysch describes a naked people who lived long lives, rarely fell sick, fought violent wars, killed and ate war captives, and had no religion or king—all statements that can be found in the *Mundus Novus* letter. Even the excessive number of years the peoples supposedly lived—150—appears in *Mundus Novus*. Ruysch refers to mountains, rivers, wild animals such as lions, and trading goods—pearls, gold, Brazilwood, and wild cinnamon (cassia)—all mentioned by Vespucci. The one missing point is the featherwork that Ruysch describes the native peoples as wearing.

Vespucci's *Mundus Novus* seems also to have influenced the *Augsburg Broadside* for the text and the visual depiction of anthropophagy in the *Augsburg Broadside* repeat many of the descriptions of Brazil in the *Mundus Novus* letter. In the *Augsburg Broadside*, the human flesh shown hanging from the beams of a house echoes the statement in *Mundus Novus*: "I saw salted human flesh hanging from house-beams, much as we hang up bacon and pork." The *Mundus Novus* letter describes the bodies of the native men and women much as they are depicted in the *Augsburg Broadside*: "They have big, solid, well-formed and well-proportioned bodies, and their complexions tend toward red . . . they also have long black hair." However, the prominent use of feathers in the image—the skirts, bracelets, and headdresses—does not appear in the *Mundus Novus*, even as the letter does refer to multitudes of parrots. Instead, in the *Mundus Novus* letter Vespucci states, "Everyone of both sexes goes around naked, covering no part of the body."[33]

Vespucci's *Lettera* has been cited as a possible source for the image of anthropophagy on *Weltkarte*. The connection lies with a story of a sailor killed and eaten in full view of his crew on third voyage. The *Lettera* did not circulate as widely as the *Mundus Novus*, even though it too was influential. First published in Florence in 1504 or 1505, it was translated into Latin, and it was the text published by Lud, Ringmann, and Waldseemüller as part of the *Cosmographiae Introductio*. Scholars have long held that the information reported is less reliable. Formisano notes that "the letter combines a first-person travel account with a scientific treatise, not without slipping into the fictional devises of an adventure yarn." The story of the sailor captured, killed, and eaten in full view of his shipmates corresponds to the voyage described by Vespucci in his familiar letter of 1502, which is the same voyage that serves as the basis for the *Mundus Novus*. But neither in the familiar letter of 1502 nor in the *Mundus Novus* does the story of the sailor appear. Because this story exists only in the *Lettera,* scholars believe it a fabrication.[34]

An edition of the *Lettera*, published by Johann Grüninger, a printer in Strasbourg, in 1509, contained woodcut images that depicted anthropophagy. One woodcut image shows a naked woman about to strike a European, while a second presents body parts being chopped on a block. The image of the European man about to be hit matches the story told in the *Lettera* of the sailor killed and eaten, whereas the second woodcut image is a more generic scene of anthropophagy in daily life. It is unlikely that the artisans who created these images had crossed the Atlantic. Rather, they freely imagined the scenes most likely from the *Lettera*. As Joan-Pau Rubiés has argued, "the artist who faithfully tried to follow a text in all its details, without the privilege of having travelled to those lands, had to rely on his iconographic prejudices." The generic scene illustrates Rubiés's point, for when the artist imagined anthropophagy, what emerged was "naked men and women cutting human limbs on a flat surface like butchers." The image conveyed anthropophagy but had nothing to do with how people lived in the Americas. Once printed, the woodcut images designed and cut by artisans in Strasbourg enhanced and visualized the text of Vespucci's *Lettera*. Once sold, read, and contemplated, the pamphlet conveyed what would be considered to be reliable knowledge about the Western Atlantic to the reading public in Germany.[35]

Yet another source of information that linked anthropophagy with the Western Atlantic were Vespucci's familiar letters written directly to his patron, which are considered by scholars to be more reliable than the published letters. Even as Vespucci makes references to Dante and Ptolemy and styles himself as a highly knowledgeable astronomer and cosmographer, the content of the familiar letters is highly descriptive. Returning sailors, sea captains, and merchants similarly would have arrived with detailed descriptions of the lands and native peoples. Because Vespucci wrote some of this down, he captured and preserved the kind of eyewitness information then in circulation. Antonello Gerbi argues that Vespucci's three familiar letters together form a corpus that is "the first coherent and deliberate description of the physical nature and inhabitants of the Brazilian coast." Such descriptions, transferred orally, would have been available to chartmakers and illuminators.[36]

In the familiar letter written in Lisbon in 1502, which recounts his voyage along the coast of Brazil, Vespucci describes anthropophagy in detail, links it to war, and provides a description of the ceremony in which prisoners of war were executed. Vespucci uses his eyewitness observer status to reassure his patron that he may be trusted and that his account is true. He writes, "And this is certain, for in their houses we found human flesh hung up for smoking, and a lot of it, and we bought ten creatures from them, both males and females, who were destined for sacrifice." Those intended for consumption were casualties of war, as Vespucci makes very clear: "And when they fight, they kill one another most cruelly, and the side that emerges victorious on the field buries all of their own dead, but they dismember and eat their dead enemies." In addition, Vespucci notes that war captives could also be imprisoned, enslaved, and incorporated into the victors' communities, with some kept "as slaves in their houses." This applied to women taken as war booty, as well as men: "If females, they sleep with them; if males, they marry them to their daughters."[37]

Vespucci did not understand the rationale behind war, but he provides an explanation purportedly learned from informants in a village where he lived for several weeks. Vespucci writes in the familiar letter: "I could not learn from them why they make war upon one another. . . . the only reason they could give was that this curse had begun 'in olden times,' and that they wished to avenge the death of their ancestors."[38] According to Vespucci, anthropophagy occurred after the war was over, and it took place in the village of the victorious side:

"At certain times when a diabolical fury comes over them, they invite their relatives and the people to dinner, and they set them out before them—that is, the mother and all the children they have got from her—and performing certain ceremonies kill with arrows and eat them; as they do the same to the aforesaid male slaves and the children that have come from them."[39]

Vespucci's account is not reliable ethnography, but he does attempt to create a cultural context for anthropophagy: it was linked to warfare and revenge, and it took place as part of a ceremony.[40]

However, the cannibalism scene painted on *Weltkarte* does not attempt to reconstruct the nature or meaning of cannibalism, even as described by Vespucci. Instead, the chartmaker (or illuminator) chose to represent anthropophagy using visual motifs familiar to Western Europeans. The human body is roasted as Europeans roast animals. Even the story of the killed and devoured sailor—as told in the *Lettera*—cannot have served as direct inspiration for the creation of anthropophagous scene that appears on *Weltkarte*. That story describes native women "hacking the Christian up into pieces" and "showing us many pieces and then eating them." The scene on *Weltkarte*—of one man turning another man impaled on a spit—does not re-create this description. Moreover, the native peoples of Brazil did not possess iron or steel, and it would be difficult to fashion from wood the implement that is shown skewering the human body. The length of the log required to support such weight would be impossibly unwieldy.[41]

The scene of cannibalism on *Weltkarte* is not presented as an illustration that characterizes local customs, as the *Augsburg Broadside* does; rather the scene stands alone. When read in conversation with other imagery placed in the Atlantic World, it advances a new argument On the *Weltkarte*, the kings of Spain and Portugal sit on their thrones, hands entwined, and gaze across the Atlantic Ocean (plate 12). And what do they see? The world chart presents them with a vast ocean extending to the west and south. Two ships are headed across the northern Atlantic, to the Spanish Indies, and two ships sail the southern Atlantic for India. The newly discovered island and mainlands in the West are painted in green—the first iconic code used to suggest the Americas. These green lands in the western Atlantic are not ruled by kings, as is the case in Africa, Europe, and India, where kings are painted on the chart. Instead, the scene of cannibalism—confirmed by the legend—depicts anthropophagy. Seen in from this perspective, the message of the world chart emerges as: the kings of Spain and Portugal must claim this land and civilize it. Given Carvajal's crest on the map, it is entirely possible that the world chart was in Rome and there seen by other high church officials—cardinals like Carvajal and even the pope. The portrayal of the continents in the western Atlantic as lands with no law, no kings, and in which sinful behavior—anthropophagy—took place argued that evangelism and colonization by the Iberian monarchs must move forward. The scene of cannibalism could also support the argument that peoples who practiced anthropophagy should be enslaved.[42]

How the horrific image of cannibalism came to be and whether Carvajal's *Weltkarte* was seen by powerful men in Rome cannot be known with certainty. Because it was a highly illuminated manuscript chart, its information (and imagery) did not circulate widely. The influence of the *Augsburg Broadside* is also difficult to pinpoint. However, when printed maps included scenes of cannibalism, these maps would spread the association between anthropophagy and the Americas to large audiences of map viewers. Waldseemüller's second world map, the *Carta Marina*, printed in 1516, is the first printed map known to include a scene of cannibalism. Where Waldseemüller placed a parrot over South America on *Universalis Cosmographia*, a brief text on *Carta Marina* notes that the peoples are anthropophagous. Farther north, human body parts hang from a tree while limbs are roasting over a fire. A textual legend, as well as labels, reinforce the message that this is a land where anthropophagous peoples live. A revised edition of the *Carta Marina*, issued in 1525 by printer Lorenz Fries after Waldseemüller's death, sold well to its intended German audience—it was reprinted in 1527 and again in 1530. According to one scholar, it became "one of the most influential representations of Indians in Renaissance map illustration."[43]

Once located on a printed map, scenes of cannibalism quickly spread to new maps. As the imagery repeated, it became better known, and as it became better known, its supposed truthfulness was reinforced. Even as the geography of the Western Atlantic became more detailed and more accurate, the vast diversity of its native peoples, landscapes, animals, and lifeways devolved into simplistic visual stereotypes. Although printed maps had short lives and were discarded when they fell out of date, the imagery on them endured. Maps had become ephemera, as we shall see in the conclusion.

Conclusion

In 1841, the Portuguese friar and later cardinal Francisco Justiniano Saraiva noted the absence of basic documents on the early Portuguese voyages of exploration in the national archives: "Of the rutters [sea logs], reports, and memoires that must necessarily had to have been written in the time of the first voyages . . . very few remain today." This absence extended to charts, leading Armando Cortesão and Avelino Teixeira da Mota to write in their monumental review of Portuguese cartography that "this almost complete disappearance of the earliest charts, which we know to have existed, is the most amazing mystery in the history of Portuguese cartography." This loss of charts is not, however, just a Portuguese story. Tony Campbell has long argued that of all *portolani*, the charts and sea logs of the Mediterranean, only a minute fraction produced in the fourteenth and fifteenth centuries have survived.[1]

Historians of historical cartography have grappled with this very question: What happened to all the maps and charts that were known to have been made? Denis Wood observes that the once ubiquitous road map is now increasingly rare: "Most of them are gone now, billions lost in the making or evaporated with the words that brought them into being. . . . Pigments have faded, the paper has rotted or been consumed in the flames. Many simply cannot be found. . . . Where have all the road maps gone: and the worlds they described?"[2]

Saraiva offered a persuasive explanation for the missing Portuguese charts: a royal policy of secrecy. One reason for their disappearance, he suggested, was the "prudent and cautious secrecy in which our rulers, at first, kept such memoires and reports." Saraiva's insight, later developed among Portuguese historians and best articulated by Jaime Cortesão, proposed that the Portuguese kings followed a policy of secrecy (*política do sigilo*) that limited the information about nautical science, geography, chartmaking, and commercial opportunities from leaving Portugal. For many years, historians of historical cartography echoed Saraiva's insight and accepted Cortesão's hypothesis, but more recently historians have come to question that secrecy can explain the missing charts. One explanation offered is the catastrophic effect of natural disasters. In particular, the Lisbon earthquake of 1755—and its subsequent tsunami and fire—is believed to have destroyed the holdings of the Casa da India, where maps and charts would have been stored. War has also been blamed: in the twentieth century, the Portuguese charts and maps that had survived outside of Portugal—in Italy and Germany—were adversely affected by the bombing campaigns and the moving of libraries and archives during World War II.[3]

Another explanation for the loss of charts and maps is that they were naturally discarded as they were replaced with new, more current versions. Tony Campbell has offered this explanation for the missing portolan charts of the Mediterranean: he notes that the practice of constantly updating the information on charts encouraged them to be treated as disposable. Once out of date or worn out, charts were thrown away, or the surface scraped clean for a new chart, or the vellum repurposed. Similarly, Alfredo Pinheiro Marques argues that, while the earthquake of 1755 certainly destroyed Portuguese charts, many had already disappeared because they were utilitarian documents regularly exposed to "maritime environments that were not very favorable to their preservation." Others agree. Likewise, Portuguese historians Maria Fernanda Alegria, Suzanne Daveau, João Carlos Garcia, and Francesc Relaño point out that "early cartographic sketches were constantly revised as new navigation opened up new horizons. Deterioration through use and the constant process of revision, must, therefore, have been the main reasons for the disappearance of the earliest Portuguese cartographic specimens."[4]

Christian Jacob has reflected on the ephemeral nature of maps, and he proposes a fundamental distinction between ephemeral and monumental maps. The ephemeral map, he writes, is "drawn on the ground, in the sand, or in ashes"; it exists in the moment, in the "gesture" of its creation. Such ephemeral maps cannot be separated from the maker who explains the map while creating it. The monumental map, on the other hand, anticipates the future, records the past, and is intended to endure. Jacob considers many kinds of monumental maps—from murals to mosaics, to those cut on stone or in wood, to those pressed into clay tablets, to those printed on paper. Even monumental maps are fragile, even somewhat ephemeral, in Jacob's view. For example, because printed maps were made with fragile materials and were repeatedly consulted, they deteriorated easily. The charts taken to sea, Jacob writes, "were often gone with the wind and the tide." Moreover, the survival of a chart or printed map depends, Jacob notes, on the "destiny of the structure that shelters it, disappearing with it in the case of fire, war, or bombardment." Jacob distinguishes further between concrete ephemeral and concrete monumental maps. The concrete ephemeral map is made using natural materials at hand—such as sand or sticks—while the concrete monumental map is made away from the location being mapped with artificial materials, such as ink on paper.[5]

These insights are all relevant to understanding how the Atlantic was first charted as an ocean and mapped as a world. Many, if not most, charts and maps have not survived, and yet their influence was profound. How is this possible? To answer this question, it is helpful to expand Jacob's definition of the ephemeral map—those made in the moment and erased immediately after having been made—and to consider *all* printed maps as ephemera and most manuscript charts as proto-ephemera. This requires setting aside the ephemeral maps fashioned and obliterated in the moment and focusing just on the kinds of charts and maps discussed in this book. When printed maps are understood as ephemera and manuscript charts as proto-ephemera, it is possible to see their far-reaching influence even as they disappear.[6]

The term "ephemera" refers to things designed without the intention of being permanent. The definition originates in the meaning of an ephemeron (plural ephemera), "an insect, which, in its winged state, lives but for a day." In its modern definition, however, the term "ephemera" describes a kind of product that lasts longer than the moment but is still intended to be discarded. For example, John Lewis defines ephemera as "printed matter of no lasting value except to collectors, as tickets, posters, greeting cards, etc." Maurice Rickards defines ephemera as the "minor transient documents of everyday life," whereas John Dann amplifies the definition to include three major elements: ephemera are produced for a practical purpose with no thought that they would be saved; ephemera fall between the cracks of traditional collecting fields and librarianship; and the diversity of ephemera documents everyday life.[7]

While printed maps and manuscript charts do not quite fit these definitions, they have much in common. Historian Catherine Delano Smith uses the term "cartographic ephemera" to refer to "maps in newssheets and pamphlets" that were "easily damaged and discarded." There is a novelty associated with ephemera, as the new edition/version replaces the old. This too operated in the past. In the sixteenth century, consumers preferred "new" maps, and artists and printers, as historian Genevieve Carlton shows, promoted their maps as "new" even when they were woefully out of date. In the modern era, printed maps have thousands of practical uses, they are diverse in nature, and they share common design elements. Because maps are temporary, utilitarian, and geographically and chronologically specific, there is a strong tendency to throw them away when they have outlived their usefulness, passed their prime, or worn out their medium. A common characteristic of print ephemera is their failure to make their way into archives, and this too explains the loss of many modern and historical maps. Complete collections of modern highway maps, for example, are rare because librarians are often uncertain of the need to catalog and preserve them. On the other hand, because the unique aesthetics and repetitive patterns of ephemera attract collectors, ephemera can become important sources for historians.[8]

The term "ephemera" captures the short lifespan of many printed maps, both historical and modern, but it does not completely describe the nature of manuscript charts. Manuscript charts were made by hand, and therefore each appears to be a unique creation. However, the ordinary charts taken to sea may have been created in multiple copies from pattern charts, and once taken on board ship, they wore out and their information became dated. The continuous production of charts constituted sets of utilitarian documents that were not deemed important enough to archive. In the sixteenth and seventeenth centuries, manuscript charts were replaced by printed charts and printed atlases. As the medium changed—from manuscript to print—printed charts (not bound into atlases) joined printed maps as examples of print ephemera.[9]

As printed maps came off the press, those subsequently bound into books tended to endure, whereas those purchased as broadsheets, or in multiple sheets to be mounted on a board or wall, tended to disappear. The printed maps created for expensive volumes of Ptolemy's *Geography,* for example, were not intended to have a short lifespan. Because these carefully printed maps were bound into scholarly books that found their way into libraries, the maps survived. Broadsheet maps had a short life span—intentionally so—and many of the extant broadsheet maps exist, Campbell suggests, only "because an early owner tipped them into one of his volumes." Similarly, large wall maps, such as Waldseemüller's world map of 1507—the *Universalis Cosmographia*—were popular, and by the middle of the sixteenth century Cornelius Koeman estimates about fifteen hundred such maps were made and mounted. However, only few such large wall maps have survived. Koeman underscores that this was due not to their marginality but to their widespread use. That comparatively few maps and charts can be studied today must not overshadow the fact that many more were made, used, and discarded. As Jacob writes, "Writing a history of cartography implies mainly studying the corpus of surviving maps," but this must not obscure "the whole cartographic production of a given period."[10]

After 1500, the number of printed maps that showed the Atlantic Ocean and that suggested the nature of the Atlantic World increased rapidly. Printers saw an opportunity: to print maps and to constantly update them by releasing new versions. Printers produced inexpensive maps aimed at "an audience of educated peoples, such as merchants," argues Susi Colin; printed maps appealed to those who "liked decorating the walls of their offices with evidence of their knowledge of geography and their world-wide trading contacts." Printed maps also began to appear in homes. Johannes Grüninger, the printer in Strasbourg, clearly saw this commercial potential of maps. Grüninger entered the map publishing business "in which he was a great success," according to historian Hildegard Binder Johnson; "he printed at least eighty thousand" large map sheets.[11]

Genevieve Carlton and Catherine Delano Smith document the maps owned by the recently deceased in sixteenth-century Italy and England. In the university town of Cambridge, Smith describes the maps recorded in inventories of the probate court of the vice-chancellor of the university, which included not only students, professors, fellows, and graduates but also university booksellers and servants. Carlton traces patterns of map ownership among elite families in Venice and Florence. Both Smith and Carlton find that maps were displayed on walls in private and public spaces and that maps and religious art were hung together. Maps seem to have been acquired in the university community for a variety of reasons, such as for study or even decoration. In Venice and Florence, maps played a role in how Venetians and Florentines presented themselves publicly as cosmopolitan, learned, and sophisticated persons.[12]

Although not every printed map from the sixteenth century constitutes ephemera, clearly many, if not the majority, were. Editions of printed maps can be seen as "sets" of ephemera where small variations in the editions or the print process slightly differentiated each map. Once printed, individual maps might be colored, highlighted, or annotated differently. The printed single-sheet maps that have endured to the present are the last remaining exemplars from such print editions. What happened to the eighty thousand sheets that Grüninger was estimated to have printed and sold? Presumably, after consumers bought, studied, and enjoyed them, most were discarded. Printers like Grüninger knew that the map they were working on today might be deemed out of date quickly, but they also knew that a new edition could easily be produced, printed, and sold in its place. The same was true in Italy, where multiple editions of the same map, slightly altered here and there, could be produced in the same year.[13]

Waldseemüller recognized the short life span of a printed map. Nine years after he published the *Universalis Cosmographia,* he changed his presentation of the world in his new world map: the *Carta Marina* (figure C.1). In the legend of the *Carta Marina*, Waldseemüller acknowledges that he knows that his previous world map has fallen out of date. The legend states that an earlier world map "was filled with error, wonder and confusion." In contrast, the legend continues, this new world map takes its information from modern nautical charts as well as the accounts of travelers, both ancient and modern. By 1516, Waldseemüller no longer had confidence that his previous world map accurately depicted the world. In the new world map, he abandoned Ptolemy's projections and used instead the layout found on the world charts made by chartmakers.[14]

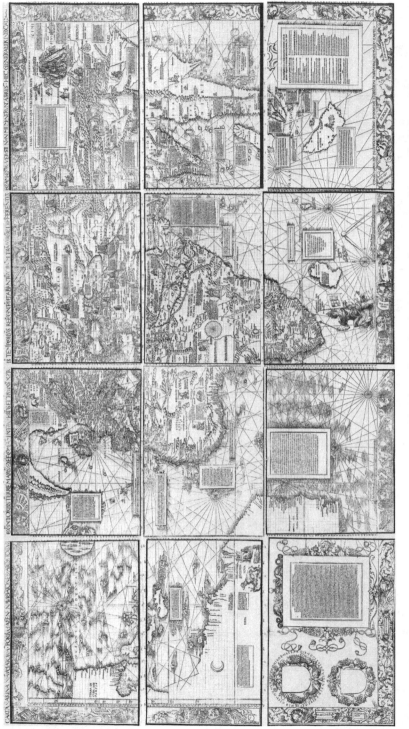

Fig. C.1 Waldseemüller's *Carta Marina*, 1516. Digitally assembled from the individual sheets of Martin Waldseemüller's *Carta Marina*, 1516, Geography and Map Division, Library of Congress, Washington, DC.

By 1516, if not earlier, Waldseemüller must have recognized that the fate of many copies of the *Universalis Cosmographia* was to be replaced. Over the years, those who owned copies of Waldseemüller's *Universalis Cosmographia* did in fact discard them. The *Carta Marina* suffered the same fate. Only one copy of each world map survives to this day, and the *Carta Marina* can be reconstructed only by combining sheets from different print runs. Monumental as both were at the moment of their creation, the *Universalis Cosmographia* and the *Carta Marina* were both ephemera.[15]

Perhaps recognizing the ephemeral nature of printed maps and texts, Johannes Schöner, a sixteenth-century mathematician and cosmographer from Nuremberg, decided to bind several texts, maps, and charts into a *Sammelband* (compilation of texts). Schöner was also a mapmaker, and this likely influenced his choice of texts—a set of the gores for his own celestial globe, a star chart now attributed to Albrecht Dürer, the twelve sheets from Waldseemüller's 1507 *Universalis Cosmographia*, and the twelve from Waldseemüller's *Carta Marina*. Schöner saved these maps and texts for his own study, but he was conscious of the need to preserve them. As Hessler notes, Schöner's bookplate pasted into the *Sammelband,* reads: "Schöner gives this to you, posterity; as long as it exists there is a monument to his spirit."[16]

The *Sammelband* became separated from Schöner's other books and entered the library of the Castle of Waldburg-Wolfegg, at Württemberg, Germany. Although it was hidden away in the library and unseen for several hundred years, historians nevertheless knew about Waldseemüller's missing 1507 world map from the *Cosmographiae Introductio* and from other sources. In 1891 historian Henry Harrisse wrote that in all likelihood Waldseemüller's world map was printed in Strasbourg, that it was very large, and that its mapping of the Americas extended to the tenth parallel, or 50° S. Ten years later, in 1901, Josef Fischer, a Jesuit historian and educator, found Waldseemüller's maps in Schöner's *Sammelband* in the library of Waldburg-Wolfegg. With Professor Franz Ritter von Wieser, Fischer published a facsimile of Waldseemüller's two world maps with commentary on them in English and German, which brought the maps to the attention of modern historians.[17]

In 2001, the Library of Congress purchased the twelve printed sheets of the 1507 *Universalis Cosmographia* that had been bound into Schöner's *Sammelband* for $10 million. The collector Jay Kislak subsequently acquired the *Sammelband*, and later donated its contents to the Library of Congress. Waldseemüller's 1507 world map, the 1516 *Carta Marina*, and the remnants of Schöner's globe gores are all on permanent display in the Library of Congress. Waldseemüller's 1507 world map is in a specially constructed case designed to withstand natural disasters. Of

Waldseemüller's globe, no original has survived, but four copies of the original printed sheets of the globe gores have endured. Recent discoveries of new sheets of the globe gores have been shown to be forgeries.[18]

If the essential medium for ephemera is print, can manuscript charts truly be considered ephemera? During the sixteenth century, the traditional methods for making charts by hand, using inks and paints on vellum, continued, but the medium of print eventually replaced the making of charts by hand. The printed chart does fit the definition of ephemera, for printed charts were utilitarian documents created for specific needs, such as a voyage, and given the environment on board ship, paper charts were not intended to last. The earlier charts are harder to classify as ephemera for not only does the traditional definition of ephemera emphasize print, but each chart, made by hand, seems to be distinctive.[19]

Manuscript nautical charts nevertheless did resemble later print ephemera in key ways. Chartmakers constantly updated their charts, which led to the discarding of older versions. Charts on vellum were prepared for specific voyages, and the chart taken to sea was a utilitarian product. Chartmakers used the same techniques and materials, the same design elements, and the shop's reservoir of geographic information repeatedly. Possibly, nearly identical charts were made in order to create portfolios of charts—some to be taken to sea and others held on land. As a result, the charts from the same shop may have been repetitive and could be seen as having been produced in sets.

The monumental world charts made using the same techniques as smaller charts can hardly have been envisioned as ephemera. Not only were they heavily illuminated and prepared for important patrons, but each was designed as a unique document, intended to last. The *Carta Universal* by Juan de la Cosa, *Carta del Cantino*, *Weltkarte*, and other manuscript world charts were documents that would not have been discarded easily. One reason why they are likely to survive, David Quinn has suggested, is because they were gifts exchanged among princes and later valued as art objects.[20]

Nevertheless, even if produced with more elaborate designs or more expensive pigments, high-end world charts still risked the same fate as printed maps and charts: to be discarded. Richly illuminated world charts that might have been given as gifts or recognized as having artistic merit still fell out of date quickly after 1500. Each chart or map that survives from circa 1500 has a singular story, and each story reveals that at key moments in time the illuminated world charts were in danger of being lost.

The story of the survival of *Carta del Cantino* is a case in point. Although it was richly illuminated, it did not survive because it was considered a work of art;

rather, it endured because it became part of the Duke of Ferrara's library and that library continues to exist today as part of the Biblioteca Estense Universitaria of Modena. Even within the library, it was nearly lost on several occasions. A fire swept through the Ferrara palace in 1554; more serious were the earthquakes of the 1570s. War was another menace. A particularly vulnerable time came when Duke Alfonso II died without a legitimate heir in 1598, and according to a papal bull, the duchy of Ferrara was to devolve to the Vatican. While Cesare d'Este hoped to resist and become the next duke, the pope deemed him to be from an illegitimate branch of the family and raised an army to march on Ferrara. An agreement, known as the Convention of Faenza, avoided war and allowed the Estes to take their moveable property to Modena, while their lands and palaces in the city of Ferrara passed into the hands of the papal court. The library, with *Carta del Cantino* still in its collection, made this transition to Modena. Once there, however, the library suffered years of neglect, and many of the artistic treasures of the Este family were lost. Eventually, in 1696, the library was properly installed in the ducal palace in Modena.[21]

Even within a stable library, the world chart was not safe. During the riots of 1859 that accompanied Italian Unification, Francesco V d'Austria-Este fled before securing his library, and several maps were stolen. *Carta del Cantino* was found in Modena, "serving as a screen at the back of a butcher's shop" by Giuseppe Boni, an art collector, who returned it to the library in 1870.[22]

When Modena was bombarded in 1944 during World War II, *Carta del Cantino* might have been in grave danger; however, it made it through the war and remains in the Biblioteca Estense Universitaria of Modena. In 2012, two serious earthquakes struck northern Italy, causing grave damage to many Italian monuments in Modena and Ferrara, and closing temporarily the archives and the Estense library, pending evaluation of the structural damage suffered. Today *Carta del Cantino* is safely stored in a specially designed glass case, and it has been recently digitized.[23]

What does seeing maps and charts as ephemera (or proto-ephemera) offer the historian of the Atlantic World? Scholars who have studied sets of ephemera emphasize that despite the fragile medium and the expectation that the finished product will be discarded, imagery on ephemera reappears repeatedly. Historian Tani Barlow argues that similar imagery reappears even as ephemera are continuously discarded. This imagery "sticks" even though individual examples of ephemera disappear. The repetition of certain images, even as the medium fades away, is revealing. The repeating images can be read for clues about social, economic, and political life, and they reveal underlying assumptions of gender roles, marketing strategies, or political campaigns.[24]

Historians of ephemera study collections, and therefore they can trace how images repeat, and they can unpack their meanings within the collection, but this approach is not feasible for historians of historical cartography. In some cases, there are no surviving examples—such as of the early Portuguese charts. In other cases, the number of surviving charts or maps is too small. Except for works in modern collections—such as the highway maps that Wood decodes for deeper meanings—historians of maps do not generally have access to reliable sets of ephemera. Nevertheless, from surviving maps and charts, historians can and must imagine the larger sets to which maps and charts belonged.

Scholars of ephemera emphasize that visual tropes repeat and that visual tropes move easily from piece to piece. In the case of manuscript charts, this can be seen in the repetition of visual codes. Even though most Mediterranean charts have been lost, historians know that the compass rose, the cityscape, and the use of color to differentiate rhumb lines were common visual codes. There are enough surviving charts to assert this. In the case of visual tropes of the Atlantic World, even though there are few surviving charts, visual imagery that first appeared on charts clearly repeated and moved from charts to maps.

Printers in cities such as Strasbourg, Augsburg, or Nuremberg could not reproduce the same layout used by chartmakers, and they often had difficulty with the textual information, such as the toponyms, but they easily adopted and simplified the visual imagery that appeared on charts. The parrot on Waldseemüller's *Universalis Cosmographia* is copied, though simplified, from the parrot on the Caverio *Planisphère nautique*. Waldseemüller (or the artisans who worked on the map) possibly modified the image of the parrot by using different print sources, and they certainly used printed images to shape the elephant. This reveals that once an image became associated with a location, additional sources of information could be used to interpret it. As accounts of travelers circulated in print, these became sources that mapmakers could use to add to the visual imagery already established by chartmakers. Just as chartmakers processed the oral knowledge of returning sailors, so too did mapmakers process the increasing amount of printed information in circulation. In this way, they influenced how the Atlantic World would appear on printed maps. For example, Waldseemüller and the artisans had many sources to use to illustrate the *Carta Marina*. In deciding how to create the scene of cannibalism, they had access to many visual representations of the Americas that circulated in print. Scenes created by mapmakers and artisans who never traveled were thus part of the defining of the Atlantic World. Once their imagery was associated with places on a map, it was "read" as reliable.

Certain images became repeating metaphors, familiar caricatures that called distant places to mind. Such was the cannibalism scene. Even as maps wore out, were disposed of, or were redesigned, the scene reappeared. Because a printed map could be produced in a long run, it disseminated textual and visual knowledge more widely than manuscript charts could. Even though print was the more modern medium, that did not make it more accurate. Printed maps not only lagged manuscript charts in accuracy, but printers could easily deceive their customers by presenting outdated information as current or even new. This repetition of visual motifs in print ephemera played a significant role in shaping the European understanding of newly found places. This is hardly a minor historical fact, for these repeating images both revealed and shaped the worldview of early modern Europeans, who, through maps, visualized the possibilities of the Atlantic World.[25]

The growing significance of maps in Europe after 1500 has been widely recognized by historians of historical cartography, leading to, in the words of David Buisseret, "a revolution in the European way of 'seeing' the world." On the sixteenth-century world maps increasingly seen in Western Europe, the Atlantic Ocean was no longer peripheral. Gerard Mercator's huge world map, published in 1569, reflects the new and prominent place of the Atlantic Ocean. In *Nova et aucta orbis terrae descriptio ad usum navigantium emendata accommodata (*A New and Enlarged Description of the Earth with Corrections for Use in Navigation*)*, the "Oceanus Atlanticus" (Atlantic Ocean) is prominently labeled, and the label is placed between the Canary Islands and the Caribbean. Farther south, there appears in the ocean a second label, "Oceanus Aethiopicus" (African Ocean), and it is placed between Brazil and West Central Africa, but closer to the African side. The Atlantic World has few visual codes, but among them are large ships and compass roses and, in the Americas, a cannibal scene and a marsupial (the possum). The cannibal scene and the possum echo imagery that appeared on Waldseemüller's *Carta Marina*, but Mercator moved them: rather than placing them in northern South America, they appear much farther south. The legend, however, attributes the custom of man-eating to native peoples living throughout the "New Indies."[26]

The significance of Mercator's world map in the history of cartography lies with the new projection, which was designed to aid navigators at sea. Although it was not yet possible in the sixteenth century to successfully navigate using charts drawn with the Mercator projection, it would come with time. The projection, which enlarges the northern and southern latitudes, dominated the way the world would be seen on maps for several centuries. The projection has been roundly criticized because it makes Europe larger and Africa and South America smaller, but the projection is still in use and underlies the making of many modern digital maps, including Google Earth.[27]

Among the historians of historical cartography, some have emphasized how maps became powerful tools of the state, and how the act of mapping legitimized some claims and erased others. Denis Wood argues that "large, centralized societies, everywhere in the world," inaugurated "mapmaking traditions as part of their transition to the early modern state." Maps served bureaucrats, military commanders, and ambitious princes, and they helped to construct the nation-state by visualizing and locating it. J. B. Harley forcefully states that maps are "preeminently a language of power" and that cartography is a discourse that reinforces the status quo and freezes social interaction. Harley points to the dark side of mapping, which can be seen in the way that military maps "not only facilitate the technical conduct of warfare, but also palliate the sense of guilt which arises from its conduct." Maps legitimized conquests and confirmed the territories of rulers.[28]

This emphasis on the state, which is an essential perspective for understanding how the Atlantic World was developed, should not obscure that in 1500 the state did not control how the Atlantic World was first recorded on charts and maps. In the years immediately following 1500, a few chartmakers, cosmographers, and artists visualized the idea of an Atlantic World. Their position is remarkable, for they situated themselves between their sources of information and their audience of viewers. They made hundreds of decisions when creating a chart or map, and they grappled with many questions explicitly and implicitly. Choices made had consequences for what the chart or map showed and how it would be read. Seemingly simple questions, such as what geographic information was reliable, were not easily answered. Even the basic layout of the chart or map—such as where the central point lay or which projection was selected—influenced how the spherical earth would appear on a flat surface, thereby shaping how the map viewer would interpret the geography and spatial relationships of the Atlantic.

Particularly during the years following 1500, when there was not yet an accepted paradigm for how to place the Atlantic Ocean within the geography of the world, or even how to represent the world on a map, those who made charts and maps had a moment of great power. Their talent as artists, their experience with making charts and maps, and their familiarity with other maps and charts— and, for some, their direct experience with the Atlantic—all influenced how they represented the Atlantic as a geographic and cultural place. For this reason, Harley's view of the cartographer as a "puppet dressed in a technical language" or his claim that "the cartographer has never been an independent artist, craftsman, or technician" does not characterize this historical moment. Rather, chartmakers and mapmakers were historical actors—often anonymous—who left a remarkable imprint on world history.[29]

In a few short years, chartmakers, cosmographers, and artists conceptualized the Atlantic, and their maps and charts would influence those who came after. This new way of seeing the world affected peoples, states, and regions that lay adjacent to the Atlantic Ocean and even beyond. As maps increasingly became important, they were nevertheless easily discarded. That so many maps became ephemera at the same time they became central to life in Western Europe is a fascinating paradox but in no way undermines the power of maps to shape how the Atlantic came to be seen.

From the first years of the sixteenth century, a few charts and maps quickly and persuasively presented a new understanding of the extent of the Atlantic Ocean. The ocean had always been there, and the ocean had not changed. Charts and maps suggested, however, that transatlantic connections, interactions, and exchanges were not only possible but desirable. The concept of the Atlantic World, which has been considered a modern idea, was expressed as early as 1500. This insight of the chartmakers and mapmakers—that there would be an inter-connected Atlantic World—is the beginning of a story that would affect millions of peoples. Many who crossed (or re-crossed) the ocean carried animals, plants, seeds, diseases, tools, beliefs, and rituals that would transform landscapes and communities. Very few of those who made the journey carried maps. Still many, if not most, understood that the formidable, dangerous, unpredictable, and violent ocean was also woven with uncountable strands—from the very personal to the most abstract—that deeply affected their own lives and of others around them. For each crossing spun new threads and broke old connections. Over distances and through time, the thick web of movement that was the Atlantic World touched everyone and everything.

Introduction

1. On the Atlantic as Ocean, see Winchester, *Atlantic*, 29–50; on the Atlantic as World, see Canny and Morgan, *The Atlantic World*, quotation from p. 1.

2. Games, "Atlantic History," quotation from 745. Rodger, "Atlantic Seafaring," 72. Edney argues that European geographers first promulgated the idea of the Atlantic Ocean and that English geographers popularized the idea of a single oceanic basin by the eighteenth century, but that the acceptance of the Atlantic probably stemmed from Britain's nineteenth-century cartographic hegemony; see "Knowledge and Cartography," 101, 111. I follow Rodger who argues that "the Atlantic world was conceptually united" by the middle of the sixteenth century; see "Atlantic Seafaring," 79.

3. See, for example, the beautiful collection of maps, organized chronologically, in Wolff, *America,* that show the increasing incorporation of the Atlantic Ocean in maps.

4. Harley and Woodward, *History of Cartography*, 1:xvi. Defining a map is extremely difficult; see Jacob's perceptive chapter "What Is a Map?," in *Sovereign Map*, 11–101; quotations from p. 30.

5. L'Anse aux Meadows National Historic Site, UNESCO World Heritage site, https://whc.unesco.org/en/list/4. On the role of the Caribbean as the staging ground for the conquest of Mexico, see Sauer, *The Early Spanish Main*. The classic account of the motivations of the conquistadores of Mexico is the memoire by footsoldier Bernal Diaz del Castillo written late in his life and published posthumously as *Historia verdadera de la conquista de la Nueva España.* Crosby's arguments appear in *The Columbian Exchange*; for Diamond see *Guns, Germs, and Steel*. Restall approaches the persistent stereotypes of the conquest in *Seven Myths of the Spanish Conquest*.

6. NOAA [National Oceanic and Atmospheric Administration], "How Big Is the Atlantic Ocean?," https://oceanservice.noaa.gov/facts/atlantic.html.

7. Edney, "Knowledge and Cartography in the Early Atlantic," 87.

8. Bailyn, "The Idea of Atlantic History," in *Atlantic History*, 3–56; quotations from p. 54; 55–56. In his 2006 review of *Atlantic History*, S. D. Smith suggests that Bailyn's account of the emergence of the field of Atlantic History is usefully understood as a paradigm shift.

9. The reviews of Bailyn's *Atlantic History* touch on many of the critiques made by historians of the concept of the Atlantic World as a frame of historical analysis. See, for example, the reviews by Timothy Coates (*e-journal of Portuguese History*, 2005), who argues

that the Atlantic World was created much earlier, by sixteenth-century chroniclers working for the Portuguese crown; Peter A. Coclanis (*Business History Review*, 2006), who recognizes the contribution of the Atlantic focus but believes the Atlantic World creates an artificial separation from the broader world. Stephen Conway similarly questions the exclusion of Asia and argues that the Atlantic and Asian worlds were not "heremetically sealed units" (*History*, 2007). Alison Games (*American Historical Review*, 2006) notes the "comparative neglect" of Africa. In "Beyond the 'Atlantic World,'" Juliana Barr argues that the Atlantic World model does not bring the four continents together equally and that it narrows the understanding of Africa and native America; see "Beyond the 'Atlantic World.'"

For the phases of development of the Atlantic World, see Bailyn's second essay in *Atlantic History*, "On the Contours of Atlantic History." The Atlantic literature is vast, but the syntheses provided in the following classic texts and textbooks are useful: Armitage and Braddick, *British Atlantic World*; Benjamin, *The Atlantic World*; Coffman, Leonard, and O'Reilly, *The Atlantic World*; D. Egerton et al., *The Atlantic World*; Elliott, *Empires of the Atlantic World*; Thornton, *Africa and Africans in the Making of the Atlantic World*; and Canny and Morgan, *The Oxford Handbook of the Atlantic World*.

For economic approaches, see the classic studies of key Atlantic commodities: Mintz, *Sweetness and Power*; S. B. Schwartz, *Tropical Babylons*; Cosner, *The Golden Leaf*; Kulikoff, *Tobacco and Slaves*; Sarson, *Tobacco-Plantation South*; and Norton, *Sacred Gifts, Profane Pleasures*. The collection of essays in Topik, Marichal, and Frank, *From Silver to Cocaine*, emphasizes the commodity chains that linked Latin American commodities to consumers, and although Topik, Marichal, and Frank propose a global economic context, the essays are clearly set in the Atlantic World. See, e.g., McCreery, "Indigo Commodity Chains"; Topik and Samper, "The Latin American Coffee Commodity Chain"; and Mahony, "The Local and the Global."

On the family in the Atlantic World, see Hartigan-O'Connor, "Marriage and Family in the Atlantic World"; Hardwick, Pearsall, and Wulf, "Centering Families in Atlantic Histories," the 2013 special issue of the *William and Mary Quarterly*; as well as Scott and Hébrard, *Freedom Papers*; Lovejoy, "The Children of Slavery"; and Manning, "Frontiers of Family Life."

For the quantitative data that underlay modern estimates of the transatlantic slave trade, see *Slave Voyages*, https://www.slavevoyages.org/. For the brutal human experience of those experiencing the slave trade, see Smallwood, *Saltwater Slavery*. For an example of the reconstruction of the experiences of one African group—the Yoruba—dispersed throughout the Atlantic World, see the essays in the anthology edited by Falola and Childs, *The Yoruba Diaspora in the Atlantic World*.

Fascinating studies of religious syncretism, hybridity, coexistence, and evangelism in the Atlantic World include Cañizares-Esguerra, *Puritan Conquistadors*; Gerbner, *Christian Slavery*; Schwartz, *All Can be Saved*; Sweet, *Recreating Africa* and *Domingos Álvares*; see also the essays in Tavárez, *Words and Worlds Turned Around*, as well as Charles, *Allies at Odds*, and Sensbach, *Rebecca's Revival*.

On the devastating population decline of native American populations, see N. Cook, *Demographic Collapse* and *Born to Die*; N. Cook and Lovell, *Secret Judgments of God*; Alchon, *A Pest in the Land*; S. Cook and Borah, *The Indian Population of Central Mexico*. On the Columbian Exchange, see Crosby, *The Columbian Exchange* and *Ecological Imperialism*. On the transformation of the landscape, see Melville, *A Plague of Sheep*; Carney and Rosomoff, *In the Shadow of Slavery*; Dean, *With Broadax*; and Metcalf, *Go-betweens*, 151–55.

On identities shaped by the Atlantic World, see Canny and Pagden, *Colonial Identity in the Atlantic World*; Sidbury, *Becoming African in America*; Garrigus and Morris, *Assumed Identities*; and Hawthorne, *From Africa to Brazil*. The scholarship on the Age of Revolution in the Atlantic World is too vast to cite here, but see Gould, *Among the Powers*; Klooster, *Revolutions in the Atlantic World*; and Paquette, *Imperial Portugal*.

10. Cañizares-Esguerra, *Entangled Empires*, 3.

11. In Atlantic World histories, the modern maps created for publication tend to be effective transmitters of knowledge, whereas the historical maps remain passive. For examples of modern maps, see map nos. 1–4 in Ferreira, *Cross-Cultural Exchange*, and maps 0.1, 0.2, 1.4, 2.2, 2.5 in Carney and Rosomoff, *In the Shadow of Slavery*. In contrast, the historical maps nos. 1.1, 3.1, 4.1, and 5.1 in *Cross-Cultural Exchange* serve mainly as illustrations as does 3.2 in Carney and Rosomoff, *In the Shadow of Slavery*. However, Carney and Rosomoff do provide a detailed reading of historical maps in order to extract information on quilombos; see nos. 5.3, 5.4, 5.5, 5.6, and 5.7. Maps in my earlier work fall into this same pattern. I began the process of unpacking the historical maps in order to use them effectively as sources in Metcalf, "Mapping the Traveled Space" and *The Return of Hans Staden*, coauthored with Eve M. Duffy.

12. Barber, "Context Is Everything . . ." An example of an illustrated carto-bibliography that documents a collection that changed over time is Kupčík, *Münchner Portolankarten/Munich Portolan Charts*, while examples of illustrated national carto-bibliographies include A. Cortesão and Mota, *Portugaliae Monumenta Cartographica*; and Schilder, *Monumenta cartographica Neerlandica*. For thematic illustrated carto-bibliographies, see, e.g., Baynton-Williams and Baynton-Williams, *New Worlds*; Betz, *Mapping of Africa*, Burden, *Mapping of North America*; Nebenzahl, *Atlas of Columbus*.

13. The *History of Cartography* project has published six volumes. Woodward coedited volumes 1and 2 with J. B. Harley, and after Harley's death, the bulk of volume3, which appeared in 2007 after Woodward's untimely passing in 2004.

14. Harley, "Deconstructing the Map"; Wood with Fels, *The Power of Maps*.

15. On the disappearance of maps, see A. Cortesão, *Cartografia portuguesa antiga*, 60–61; Marques, *Origem e desenvolvimento*, 85; Wood with Fels, *The Power of Maps*, 4; Wood, Fels, and Krygier, *Rethinking the Power of Maps*, 15; and T. Campbell, "Portolan Charts from the Late Thirteenth Century," 372–73; and T. Campbell and Destombes, *The Earliest Printed Maps*, 16. On maps surviving because they were deemed works of art, see Quinn, "Artists and Illustrators."

16. For example, the National Ocean Service establishes the modern difference as follows: "A nautical chart represents hydrographic data, providing very detailed information on water depths, shoreline, tide predictions, obstructions to navigation such as rocks and shipwrecks, and navigational aids. The term "map," on the other hand, emphasizes landforms and encompasses various geographic and cartographic products. . . . A map usually represents topographical information"; NOAA, "What Is the Difference between a Nautical Chart and a Map" https://oceanservice.noaa.gov/facts/chart_map.html.

17. *Universalis Cosmographia Secundum Ptholomaei Traditionem et Americi Vespucii Alioru[m]que Lustrationes*, St. Dié, 1507, Library of Congress, https://www.loc.gov/rr/geogmap/waldexh.html.

Chapter 1 · The Atlantic on the Periphery

1. On Heracles erecting the pillars at "the furthest limit of voyaging," see "The Nemean Odes III 18–35," in Sandys, *Odes of Pindar*, 336. The ancient city of Gades (later Cadiz) lay beyond the straits on the island of Erytheia; see "Gades," *Ancient History Encyclopedia*, https://www.ancient.eu/Gades/. On the Pillars/Columns of Hercules on maps, see Relaño, *Shaping of Africa*, 149–60.

2. In a subsequent, unfinished dialogue, the *Critias*, Plato gives a few more details, among them that the island of Atlantis had been given to Poseidon, that his children inhabited it, and that it was named after his son Atlas. Plato, *Timaeus*, 24e–25d; Gill, *Plato's Atlantis Story*, 66–67; 77–78; Zeyl, introduction in Plato, *Timaeus*, xxvii–xxviii.

3. On the loss of maps from the classical period, see Dilke, "Cartography in the Ancient World," 106. On Anaximander, see Dilke, *Greek and Roman Maps*, 22–23, and Aujac, "Foundations of Theoretical Cartography," 134–135. Herodotus's remarks about maps from Dilke, *Greek and Roman Maps*, 24–25. Aristotle believed the earth to be a sphere divided into five zones, only two of which were fit for humans. Geographically, this zone extended from the Straits of Gibraltar to India and excluded the far north and the far south; see Dilke, *Greek and Roman Maps*, 27–29, and Aujac, "Foundations of Theoretical Cartography," 135, 145.

4. "In totum abstulit terras primum omnium ubi Atlanticum mare est, si Platoni credimus, immenso spatio, mox interno quoque quae videmos hodie," Pliny, *Natural History*, 2:92(205) from *Pliny Natural History*, trans. H. Rackham, 1:334–36. Romans knew of the story of Atlantis from translations of the *Timaeus* dialogue from Greek to Latin. Many scholars reason, as Pliny may have, that Plato invented the story of Atlantis for his own literary and philosophical ends. "Terrarum orbis universus in tres dividitur partes, Europam Asiam Africam. origo ab occasu solis et Gaditano freto, qua inrumpens oceanus Atlanticus in maria interiora diffunditur. hinc intranti dextera Africa est, laeva Europa, inter has Asia; termini amnes Tanais et Nilus," Pliny, *Natural History* 3:1(2–5) from *Pliny Natural History*, trans. Rackham, 1:3–5. Dilke argues that Roman geographical manuscripts "are of less cartographic interest than their Greek counterparts" and notes that it is unclear if Roman authors wrote with maps in front of them or if maps were drawn to illustrate their texts. In the case of Pliny, Dilke writes, "it is clear that Pliny was a user of maps rather than a contributor to the theory of their construction and compilation." See "Itineraries and Geographical Maps," 242–43.

5. On Marinus of Tyre, see Dilke, "The Culmination of Greek Cartography," 178–80; on the equirectangular projection, see Snyder, *Flattening the Earth*, 2, 5–6. On Ptolemy, see Dilke, "The Culmination of Greek Cartography," and for a modern translation, Berggren, Jones and Ptolemy, *Ptolemy's Geography*.

6. *Mappaemundi* is defined by David Woodward as "from the Latin *mappa* (a tablecloth or napkin) and *mundus* (world)," but as Woodward notes, in medieval Europe, the term could also refer to textual descriptions; Quotation on time and space from Woodward, "Reality, Symbolism, Time, and Space," 511. On the number of surviving *mappaemundi*, see Woodward, "Medieval *Mappaemundi*," 286. For *mappaemundi* in manuscripts, see the extensive commentary in Pérez, *The Beatus Maps*. Woodward identifies the four basic categories of *mappaemundi* as tripartite, zonal, quadripartite, and transitional in "Medieval *Mappaemundi*," 295–99.

7. Isidore of Seville (560–636) was born in Visigothic Spain and became bishop of Seville in 603. An orator, teacher, and writer, he left his *Etymologies* unfinished at his death, and it was completed by the bishop of Zaragoza, Braulio; see Manuel C. Díaz y Díaz, "Isidore of Seville," in *Encyclopedia of Religion*, 7:4556–57. This *mappamundi* appears in the edition of Isidore of Seville's *Etymologiae*, printed in Augsburg by Gunther Zainer in 1472. For a digital copy of the edition held in the Bayerische Staatsbibliothek in Munich, see https://app.digitale-sammlungen.de/bookshelf/bsb00029495; for the actual map, see liber xiii: https://api.digitale-sammlungen.de/iiif/image/v2/bsb00029495_00378/full/full/0 /default.jpg. This T-O map is well studied; see S. Edgerton, "From Mental Matrix," 27; Edson, *The World Map*, 12–13; Nebenzahl, *Mapping the Silk Road*, 26–27; S. Davies, *Renaissance Ethnography*, 29; Woodward, "Medieval *Mappaemundi*," 301–2.

8. "Oceanum Graeci et Latini ideo nominant eo quod in circuli modum ambiat orbem," Isidore de Seville, *Etymologies*, bk. XIII, xv; "Undique enim Oceanus circumfluens eius in circulo ambit fines," Isidore de Seville, *Etymologies*, bk. XIV, ii; see also the translations by Barney et al., *Etymologies*, 277, 285. In *De natura rerum*, a schoolbook written by Isidore of Seville, he states "the ocean, spread out on the peripheral regions of the globe, bathes the borders of almost the whole world" (Oceanus autem regione circumductionis sphaerae profusus, prope totius orbis adluit fines). A slightly different translation is given by Woodward: "The ocean, spread out on the peripheral regions of the globe, bathes almost all the confines of its orb"; see "Medieval *Mappaemundi*," 320, n. 162. On Isidore of Seville's conception of the earth, see W. Stevens, "The Figure of the Earth."

9. "Quique a proximis regionibus diversa vocabula sumpsit: ut . . . Athlanticus," and "Nam Gaditanum fretum a Gadibus dictum, ubi primum ab Oceano maris Magni limen aperitur; unde et Hercules cum Gadibus pervenisset, columnas ibi posuit, sperans illic esse orbis terrarum finem," Isidore de Seville, *Etymologies*, bk. XIII, xv.

10. Woodward, "Medieval *Mappaemundi*"; quotation from Woodward, "Reality, Symbolism, Time, and Space," 514. On the placement of monsters and the Garden of Eden on *mappaemundi*, see Edson, *The World Map*, 11–32; Woodward, "Medieval *Mappaemundi*"; Scafi, *Mapping Paradise*, 125–131; S. Davies, *Renaissance Ethnography*, 29.

11. *Mappa Mundi*, Hereford Cathedral, https://www.themappamundi.co.uk/index.php. The original language of the text, which gives the mapmaker's name, is Anglo Norman, the dialect of French spoken in England by the educated classes following 1066: "De Richard de haldingham o de Lafford . . . Ki la fet e compasse," as transcribed in Harvey and Westrem, *Hereford World Map*, 1:86, and Harvey and Westrem, *Hereford World Map*, 2:xxi–xxiii. Westrem describes the map as "a sophisticated textual and pictorial representation of the world" (xv) and makes an analogy to a tapestry "carefully woven out of information derived principally from ten ancient and early medieval sources" (xxviii), among which are the Bible, Pliny the Elder, the Antonine Itinerary, Solinus, Orosius, and Isidore of Seville (xxviii–xxxvii). For a description of the map see Harvey and Westrem, *Hereford World Map*, 1:9–17; for its sources, 67–84.

12. Harvey and Westrem, *Hereford World Map*.

13. "Calpes et Abbina Gades Herculis esse creduntur," Harvey and Westrem, *Hereford World Map*, 2:426–27. The pillars, or columns, referred to two promontories on each side of the Strait of Gibraltar—Calpé on the northern side and Abyla on the southern, that Hercules created, according to Greek legend; see "Hercules, n." *OED Online*,

https://www-oed-com.ezproxy.rice.edu/view/Entry/86115?redirectedFrom=hercules.
Historian G. R. Crone notes that the legend of the Columns (or Pillars) of Hercules is
common to medieval works, writing that the columns stood at the "western limit of his
explorations, and it gradually came to be thought that these marked the limits of possible
navigation in the west"; see Crone, "Origin of the Name Antillia," 261.

14. Harvey and Westrem, *Hereford World Map*, 2:372–73.

15. T. Campbell, "Portolan Charts," quotation from 372.

16. [Abraham Cresques], *Atlas de cartes marines*, Bibliothèque nationale de France,
Département des Manuscrits, Espagnol 30, https://gallica.bnf.fr/ark:/12148/btv1b55002481n.

17. The compass rose and the rhumbs are only partially integrated. The compass rose
only intersects with the rhumb lines for the four major directions N, E, S, W. If it had been
drawn over the central point of an invisible circle, all of its thirty-two directional points
would correspond to rhumb lines. For transcriptions of the legends, see Ceva, *The Cresques
Project*, Catalan Atlas Legends, Panel III.5, http://www.cresquesproject.net/catalan-atlas
-legends/panel-iii.

18. Relaño, *Shaping of Africa*, 94–95; "Canárias, Ilhas," in Albuquerque and
Domingues, *Dicionário de história*, 1:187–89; Abulafia, *Discovery of Mankind,* 33–101; Rubiés,
"The Worlds of Europeans," 23–24; and Soule, "From Africa to the Ocean Sea."

19. "Açores," in Albuquerque and Domingues, *Dicionário de História*, 1:12–15.

20. "Madeira," in Albuquerque and Domingues, *Dicionário de História*, 2:637–39;
"Açúcar," in Albuquerque and Domingues, *Dicionário de História*, 1:15–19. On the early
sugar plantations on the Atlantic islands, see Rau, *O açúcar da Madeira*; and S. B.
Schwartz, *Sugar Plantations*, 7–15.

21. Given as "tembuch" (Timbuktu), "geugeu" (Gao), and "mayma" (Niamey), in
Relaño, *Shaping of Africa*, 97.

22. Games emphasizes this point, arguing that the land masses surrounding the Atlantic
Ocean were varied in their microclimates, disease environments, languages, and isolation;
see "Atlantic History," 742. On parallels to the Mediterranean World, see Wigen,
"Introduction: AHR Forum Oceans of History"; and Horden and Purcell, "The
Mediterranean and 'the New Thalassology.' "

23. Woodward, "Medieval *Mappaemundi*," 299; Johnson, *The German Discovery*, 51–57;
Berggren, Jones, and Ptolemy, *Ptolemy's Geography,* 4; Shalev, "Main Themes," in Shalev
and Burnett, *Ptolemy's Geography*, 1–3.

24. *Mappamondo Catalano estense*, Biblioteca Estense Universitaria, Modena
https://www.gallerie-estensi.beniculturali.it/opere/carta-catalana/. Milano and Batini,
Mapamundi Catalán Estense. Most probably made in Majorca in the middle of the fifteenth
century, the *Mappamondo Catalano estense* was acquired by the Este family in Ferrara, where
it became part of the ducal library.

25. *Il Mappamondo di Fra Mauro*, 1450, Biblioteca Marciana di Venezia, https://
marciana.venezia.sbn.it/la-biblioteca/il-patrimonio/patrimonio-librario/il-mappamondo-di
-fra-mauro. Fra Mauro was a brother at the Camaldolese monastery of San Michele di
Murano. The original map is located today in the Biblioteca Nazionale Marciana, Venice.
For a transcription and detailed study of the map, see Falchetta, *Fra Mauro's World Map*,
and Cattaneo, *Fra Mauro's Mappa Mundi*. On the sources for the map, see Falchetta, *Fra
Mauro's World Map*, 33–69; Cattaneo, *Fra Mauro's Mappa Mundi*, 118–23, 185–225, 242–44.

On the Portuguese charts, see Cortesão and Mota, *Portugaliae Monumenta Cartographica*, 1:xiii, xxxi, xxxiv, xlv; and Cattaneo, *Fra Mauro's Mappa Mundi*, 19–20. Ptolemy's work was known in Venice by 1423, and the Venetian mapmaker Andrea Bianco had drawn a Ptolemaic map in 1436; see Andrea Bianco, *L'Atlante*, Biblioteca Nazionale Marciana, Venezia. Ms. It. Z, 76 (= 4783), https://www.movio.beniculturali.it/bnm/ridottiprocuratorisanmarco/it/118/andrea-bianco. For a discussion of Fra Mauro's consideration and ultimate rejection of Ptolemaic mapping, see Falchetta, *Fra Mauro's World Map*, 52–55; and Cattaneo, *Fra Mauro's Mappa Mundi*, 376–85. Cattaneo observes that "an analysis of the legends in which the Camaldolese friar criticizes, corrects, and generally reflects upon the *Geography* clearly demonstrates the profound relationship between the two works"; *Fra Mauro's Mappa Mundi*, 165. On Mauro's rejection of Ptolemy's grid, because it would not place Jerusalem at the center of the world, see Carlton, *Wordly Consumers*, 38–39; and Scafi, *Mapping Paradise*, 238.

26. "Navegação primeira de Luiz de Cadamosto," in Academia Real das Sciencias, *Collecção de Noticias*, 2:18.

27. "Illa de Cades, asi posa Ercules does colones," in Milano and Battini, *Mapamundi Catalán Estense*, 192. A similar pair of islands and legend appears on Mecia de Viladestes, *Carte marine*, 1413, Bibliothèque nationale de France, Cartes et plans, GE AA-566 (RES), https://gallica.bnf.fr/ark:/12148/btv1b55007074s.

28. C. Riley, "Ilhas Atlânticas e Costa Africana," in Bethencourt and Chaudhuri, *História da expansão portuguesa*, 1:157–58; "Navegação segunda de Luiz de Cadamosto," in Academia Real das Sciencias, *Collecção de Noticias*, 2:59.

29. "Io ho più volte aldido da molti che qui è uma colona cum uma man che dimostra cum scriptura che de qui non se vadi più avanti. Ma qui voglio che portogalesi che navegano questo mar dicano se l'è vero quel che ho audito perché io non ardiso affermarlo," as transcribed and translated by Falchetta, *Fra Mauro's World Map*, 295–96.

30. "Nota che le colone de Hercules non significa altro che division di monti, i qual come dice le fabule seravano el streto di zibelterra," as transcribed and translated by Falchetta, *Fra Mauro's World Map*, 425.

31. Shalev, "Main Themes," in Shalev and Burnett, *Ptolemy's Geography*, 2–4. Jacopo d'Angelo da Scarperia is also known as Iacopo Angeli, Jacobus Angelus, and Giacomo d'Angelo da Scarperia, as well as other variations.

32. Dalché, "Reception of Ptolemy's Geography," 287–309; Shalev, "Main Themes," in Shalev and Burnett, *Ptolemy's Geography*, 7.

33. Quotations from Ptolemy's *Geography* from the translation by Berggren, Jones, and Ptolemy, *Ptolemy's Geography*, 57, 64. Berggren and Jones translate *geōgraphia* as cartography (a term coined only in the nineteenth century) in order to convey the sense of world mapping; see Berggren, Jones, and Ptolemy, *Ptolemy's Geography*, 57, n. 1. On Ptolemy, see Dilke, "The Culmination of Greek Cartography"; Dalché, "Reception of Ptolemy's Geography"; and Brotten, *A History of the World*, 17–53.

34. This manuscript edition of Ptolemy's *Geography* is held by the Biblioteca Estense Universitaria, Lat.463+alfa.X.1.3, and it is known by various names: *Cosmografia de Tolomeo*, *L'atlante di Borso d'Este*, or *Cosmografia*. We shall cite it as Ptolemy, *Cosmographia*. A digital copy is available on line at http://bibliotecaestense.beniculturali.it/info/img/geo/i-mo-beu-alfa.x.1.3.html. For Germano's world map, known as *Planisfero*, see also the digital

version posted at the Gallerie Estensi, https://www.gallerie-estensi.beniculturali.it/opere
/cosmografia-di-tolomeo/. A beautiful facsimile edition was published as *L'atlante di Borso
d'Este* by Il Bulino in 2006; this edition includes the text and all of the maps.

35. Berggren, Jones, and Ptolemy, *Ptolemy's Geography*, 57–59.

36. Nicolò Germano, *Planisfero*, in Ptolemy, *Cosmografia*, Biblioteca Estense Universitaria, Modena, available at https://www.gallerie-estensi.beniculturali.it/opere/cosmografia
-di-tolomeo/; quotation from Berggren, Jones, and Ptolemy, *Ptolemy's Geography*, 86.

37. *Descriptio Arcipelagi et Cicladum Aliarorum Insularum* (*Description of the Archipelago,
the Cyclades, and the Other Islands*) and also titled *Liber insularum Archipelagi*; some
seventy manuscripts from the four editions survive today. It was the first book of islands to be
illustrated by maps. See Van Duzer, *Henricus Martellus's World Map*, 4–18, and the facsimile
edition Buondelmonti, Germanus, and Edson, *Description of the Aegean*, 1–11.

38. Henricus Martellus, *Insularium illustratum*, Add MS 15760, British Library,
available at Pelagios Project https://data.bl.uk/pelagios/pelo2.html. For the mappamundi
only, https://www.bl.uk/collection-items/world-map-by-henricus-martellus. That Martellus might have seen charts that mapped the voyage of Bartolomeu Dias is possible given
that Columbus, or his brother Bartholomew, saw such a chart, or charts, in Lisbon, as
revealed in postil 490 of Columbus's copy of Pierre d'Ailly's *Ymago mundi*; see Wey
Gómez, *Tropics of Empire*, 134, n. 92, and 154–55. On Martellus's *mappamundi* map, see
A. Davies, "Behaim, Martellus and Columbus"; Edson, *The World Map*, 216–219; and Van
Duzer, *Henricus Martellus's World Map*, 16–18.

39. On the padrões, see Ravenstein, "Voyages of Diogo Cão and Bartholomeu Dias."
Padrões appeared on other charts, such as *Carta del Cantino*. See chapter 5.

40. Hec est uera forma modern affrice secundum desr[i]ptionem Portugalensium inter
mare Mediterraneaum et oceanum meridionalem," as transcribed and translated by Van
Duzer, *Henricus Martellus's World Map*, 16. The second legend describes the voyage of Diogo
Cão and the erection of the padrões—an illustration of one also appears on the map. The
third legend refers to the voyage of Bartolomeu Dias. See Van Duzer, *Henricus Martellus's
World Map*, 17–18.

41. The Martellus world map, known as *Map of the World of Christopher Columbus*, is
in the Beinecke Rare Book and Manuscript Library, Yale University, https://brbl-dl
.library.yale.edu/vufind/Record/3435243. Many texts and legends were illegible until Chet
Van Duzer secured a grant from the National Endowment for the Humanities to use
multispectral imaging to photograph the map in 2014. For the stunning results, and
discussion on the toponyms of the African coast, see Van Duzer, *Henricus Martellus's
World Map*, 126–34. On Vasco da Gama, see Subrahmanyam, *Career and Legend*.

42. Behaim-Globus, Germanisches National Museum, https://www.gnm.de/index.php
?id=749&L=1. The classic printed facsimile is Ravenstein, *Martin Behaim*, now available
georeferenced in GoogleEarth in the David Rumsey Historical Map Collection; see
https://www.davidrumsey.com/luna/servlet/detail/RUMSEY-8-1-291872-90063412:Compo
site--Times-Projection--Martin. For the globe in figure 1.8, we used the black and white
drawing in Paullin and Wright, *Atlas of the Historical Geography*, pl. 8. On Behaim's globe,
see Van Duzer, *Henricus Martellus's World Map*, 27, 172–74; Ravenstein, *Martin Behaim*,
quotation from 59–60; Goerz and Scholz, "Semantic Annotation"; and Menna et al., "High
Resolution 3D Modeling of the Behaim Globe."

Chapter 2 · The Year 1500

1. Quotation attributed to Columbus from Las Casas: "Tambien, señores Príncipes . . . tengo propósito de hacer carta nueva de navegar, en la cual situaré toda la mar é tierras del mar Océano en sus próprios lugares, debajo de su viento"; Las Casas, *Historia de las Indias*, 1:XXXV, 263. Wey Gómez, *Tropics of Empire*, 17, 119–203. The *Carte de Christophe Colomb* was attributed to Columbus by Charles de la Roncière in 1924, but it has never been confirmed; see Nebenzahl, *Atlas of Columbus*, 22–25; for the sketch, 26. Cosa, *Carta universal de Juan de la Cosa*, 1500, MNM-257, Museo Naval de Madrid & Biblioteca Virtual del Ministerio de Defensa, Madrid, http://bibliotecavirtualdefensa.es /BVMDefensa/i18n/consulta/registro.cmd?id=16822. The *Carta del Cantino*, also known as *Charta del navicare per le isole novamente trovate in la parte dell'India*, is cataloged as C. G. A. 2, Biblioteca Estense Universitaria, Modena, and available online in the Gallerie Estensi: https://www.gallerie-estensi.beniculturali.it/opere/carta-del-cantino/. Wald-seemuller's world map, *Universalis Cosmographia Secundum Ptholomaei Traditionem et Americi Vespucii Alioru[m]que Lustrationes,* is on permanent display in the Jefferson Building at the Library of Congress; see https://www.loc.gov/rr/geogmap/waldexh.html.

2. Wey Gómez, *Tropics of Empire*, 11. Van Duzer argues that Columbus's thinking fit into not only Behaim's representation of the Atlantic but also Martellus's *mappamundi* of 1491; see *Henricus Martellus's World Map*. The map by Paolo de Toscanelli, which was believed to have influenced Columbus, is so shrouded in controversy that we shall not consider it here; for that controversy, see Van Duzer, *Henricus Martellus's World Map*, 181–91. For a description of the caravel, a boat rigged for Atlantic sailing, especially along the coasts and up the rivers of West Africa, see "Caravela," in Albuquerque and Domingues, *Dicionário*, 1:197–99. The oath taken by La Cosa is recorded as "Johan de la Cosa . . . dijo . . . que nunca oyó ni vido isla que pudiese tener trescientas treinta y cinco leguas en una costa de Poniente á Levante . . . y que ciertamente no tenia dubda alguna que fuese la tierra-firme" in Navarrete, *Colección de los viages* (1825), 2:146. Quotation from Las Casas: "Colon, por quien la divina Providencia tuvo por bien de descobrir aquesta nestra grande tierra firme," in Las Casas, *Historia de las Indias*, 2:268. On Columbus's last voyage, see Wey Gómez, *Tropics of Empire*, 29–36.

3. *Carta universal de Juan de la Cosa*, 1500. Quotation from Las Casas: "el mejor piloto que por aquellas mares habia, por haber andado en todos los viajes que habia hecho el Almirante," in Las Casas, *Historia de las Indias*, 3:10. Quotation from Anghiera: "tanti momenti virum, ut Regio diplomate navium magister appellatus fit regius," in Anghiera, *De Orbe Novo* (1516), Second Decade, Chapter Seven; a slightly different emphasis is given by Gaffarel: "cet illustre navigateur que avait reçu le diplôme de pilote royal," in Anghiera, *De Orbe Novo*, trans Gaffarel, 191, which is adopted by MacNutt: "that illustrious navigator who had received a royal commission as pilot," in *De Orbe Novo*, trans. MacNutt, 1:70. The description of the Bishop of Burgos's collection of maps is in the Second Decade, Book Ten, where the Latin words for the globes and maps are "solidam universi cum his inventis spheram & membranas: quas nautae chartas vocant navigatorias." For the English translation, see Anghiera, *De Orbe Novo*, trans. MacNutt 1:271–72. On the possiblility that La Cosa's *Carta Universal* was made for Ferdinand and Isabela, see Martín-Merás, "La Carta de Juan del la Cosa," 79. On the possibility that it may have been made by a group, see Silió, *Carta de Juan de la Cosa*, 245–246.

4. Silió, *Carta de Juan de la Cosa;* Martín-Merás, "La Carta de Juan del la Cosa," 74; Gaspar, "Revistando o planisférico de Juan de la Cosa," 3–5.

5. "Juan de la Cosa la fizo en el Puerto de Santa Maria en año de 1500."

6. Text of first legend over southeast Africa: "Hasta aqui descubrio el excelente Rey don Juan Rey de Portugal"; text of second legend over India: "Tierra descubierta por el Rey dom Manuel Rey de Portugal."

7. Vespucci was born in Florence in 1452 or 1454, the son of ser Nastagio di Amerigo Vespucci, a notary. He entered the service of Lorenzo di Pierfrancesco de' Medici in the 1480s and by 1492 was in Seville. This letter, dated 18 July 1500, is available as "Nota d'una lettera scrive Amerigo Vespucci di Cadisi di loro ritorno de l'isole d'India, come apresso; e prima" in the critical edition of Vespucci's letters by Luciano Formisano, *Lettere di viaggio*, 3–14. The English translations are by David Jacobson from Vespucci and Formisano, *Letters from a New World*, 3–18. Quotation from Vespucci's letter: "E la presente sarà per darvi nuova come circa d'uno mese fa che venni delle parti della India per la via del mare Oceano con la grazia di Dio a salvamento a questa città di Sibilia [Sevilla], e perché credo che Vostra Magnificenza arà piacere di intendere tutto el successo del viaggio e delle cose che più maravigliose mi si sono offerte," in Vespucci and Formisano, *Lettere di viaggio*, 3; English translation from Vespucci and Formisano, *Letters from a New World*, 3. On the Ojeda expedition, see Navarrete, *Colección de los viages* (1829), 3:3–11; Pohl, *Amerigo Vespucci*, 51–75; Fernández-Armesto, *Amerigo*, 62–67. Formisano, *Letters from a New World*, xxv, notes that "Vespucci is said to have served as 'pilot' (that is, as astronomer and cartographer)," but Fernández-Armesto, following Las Casas, sees him more as "an adventurer." Vespucci had been living in Seville and working as an agent of Gianotto Berardi, the financial backer and outfitter of Columbus, when he joined the expedition led by Alonso de Ojeda; see Fernández-Armesto, *Amerigo*, 47–65. Las Casas describes Vespucci as a "mercador" (merchant) who invested in the armada "por aventura"; see *Historia de las Indias*, 2:271. At the end of the expedition, 232 Indians were captured in the Bahamas, of which 200 survived the voyage to Spain, where they were sold as slaves in Cadiz, as recounted by Vespucci in the letter to de' Medici; see Vespucci and Formisano, *Lettere di viaggio*, 12–13; and Vespucci and Formisano, *Letters from a New World*, 15–16.

8. "E se io sono alcuno tanto prolisso, pongasi a leggerla quando più d'ispazio estarà" in Vespucci and Formisano, *Lettere di viaggio*, 3; English translation from Vespucci and Formisano, *Letters from a New World*, 3. "Ho acordato, Magnifico Lorenzo, che così come vi ho dato conto per lettera dello che m'è occorso, mandarvi dua figure della discrezione del mondo fatte et ordinate di mia propria mano e savere: e sarà una carta in figura piana et uno apamundo in corpo sperico," in Vespucci and Formisano, *Lettere di viaggio*, 13; English translation from Vespucci and Formisano, *Letters from a New World*, 17. Vespucci stated that Francesco Lotto, from Florence, would take the map and globe to de' Medici by sea, which might mean that he was a sea captain or merchant: "el quale intendo di mandarvi per la via di mare per un Francesco Lotti, nostro fiorentino, che si truova qui," in Vespucci and Formisano, *Lettere di viaggio*, 13.

9. Anghiera's observation: "Americus vespucius florentinus vir in hac arte peritus"; Anghiera, *De Orbe Novo* (1516) Second Decade, Chapter Ten; for English, see Anghiera, *De Orbe Novo*, trans. McNutt, 1:271–72. Statement by Las Casas: "Porque como Américo era latino y elocuente," *Historia de las Indias*, 2:268; what Las Casas meant by "latino" is not

clear—possibly that Vespucci was Italian or that he was well versed in Latin. Las Casas was particularly angry that Vespucci's published letters, the *Mundus Novus* (1504) and the *Lettera* (1505), gave credit to Vespucci, not Columbus, for the discovery of the continent of South America; see Las Casas, *Historia de las Indias*, 2:268–74. The quotation by Fernández-Armesto is from Fernández-Armesto, *Amerigo*, ix.

10. Quotations from Vespucci: "Credo che vi contenteranno, e *maxime* il corpo sperico, ché poco tempo fa che ne feci uno per l'Altezza di questi re, e lo stimòn molto," in Vespucci and Formisano, *Lettere di viaggio*, 13; English translation from Vespucci and Formisano, *Letters from a New World*, 17. "Perché, come vedrete per la figura, la lor navigazione è di continuo a vista di terra, e volgono tutta la terra d'Africa per la parte d'austro," in Vespucci and Formisano, *Lettere di viaggio*, 14; English translation from Vespucci and Formisano, *Letters from a New World*, 17. "Perché mia intenzione era di vedere se potevo volgere uno cavo di terra che Ptolemeo nomina il Cavo di Cattigara, che è giunto con el Sino Magno, ché secondo mia opinione non stava molto discosto d'esso," in Vespucci and Formisano, *Lettere di viaggio*, 4, English translation from Vespucci and Formisano, *Letters from a New World*, 4. The Sinus Magnus or Sino Magno referred to by Vespucci corresponds to the Magnus Sinus on the Ptolemaic map of Asia, known today as the Bay of Bengal; see "Undecima Asie tabula," in Ptolemy, *Cosmographia* (1486), available online at https://purl.stanford.edu/fs844yc9264.

11. Quotations from Vespucci: "E tanto navigammo per la torrida zona alla parte d'austro . . . e del tutto perdemmo la stella tramontana," in Vespucci and Formisano, *Lettere di viaggio*, 5; English translation from Vespucci and Formisano, *Letters from a New World*, 6. "Non manca in cotesta città chi intenda la figura del mondo e che forse emendi alcuna cosa in essa; tuttavolta, chi mi dé emendar, aspetti la venuta mia, che potrà essere che mi difenda," in Vespucci and Formisano, *Lettere di viaggio*, 14; English translation from Vespucci and Formisano, *Letters from a New World*, 17. For Vespucci's appointment as *Piloto Mayor*, see "Real título de Piloto mayor, con extensas facultades, á Amérigo Vespucio," in Navarrete, *Colección de los viages* (1945), 3:299–302, available in English as "Appointment of Amerigo Vespucci as Chief Pilot [1508]," in Vespucci, *Letters from a New World*, 107–12. On the royal pattern chart, see Martín-Merás, *Cartografía Marítima Hispana*, 72–74; Alison Sandman, "Mirroring the World," in P. Smith and Findlen, *Merchants and Marvels*, 85–91; and Sandman, "Spanish Nautical Cartography," 1107–11.

12. Quotation from Vespucci: "et ora nuovamente il re di Portogallo tornò d'armare XII navi com grandissima ricchezza e l'ha mandate in quelle parte, e certo ch'e e' faranno gran cosa," in Vespucci and Formisano, *Lettere di viaggio*, 14; English translation from Vespucci and Formisano, *Letters from a New World*, 18. King Manuel's letter to the zamorin is reproduced as "Carta de D. Manuel ao Samorim de Calicut, 11 de Março de 1500," in Amado and Figueiredo, *Brasil 1500*, 63–72. Three letters survive from this landing in Brazil: "Carta de Pero Vaz de Caminha, 1º de Maio de 1500," "Relação do Português Anônimo, 1500," and "Carta de Mestre João, 1º de Maio de 1500," available in Amado and Figueiredo, *Brasil 1500*, 73–122, 131–41, and 123–30, respectively. For English versions, see Greenlee, *The Voyage of Pedro Alvares Cabral*. Vespucci apparently wrote about the return of the supply ship, for he makes a reference to a letter written on 8 May 1501, but this letter has not survived; see Vespucci and Formisano, *Letters from a New World*, 19, n. 1.

13. The letter from Cape Verde is dated 4 June 1501 and is available in Vespucci and Formisano, *Lettere di viaggio*, 15–20, as "Copia d'una lettera scritta 'Amerigo Vespucci de

l'isola del Capo Verde e nel mare Oceano a Lorenzo di Piero Francesco de' Medici sotto di IIII di giugno 1501.'" The English translation is from Vespucci and Formisano, *Letters from a New World*, 19–27. Quotation from Vespucci: "co' quali i' ho aùto grandissimi ragionamenti, non tanto de loro viaggio, come della costa della terra che corsone, e delle richezze che trovorono, e di quello che tengono," in Vespucci and Formisano *Lettere di Viaggio*, 15; English translation from Vespucci and Formisano, *Letters from a New World*, 20.

14. Quotation from Vespucci: "posono in una terra, dove trovonno gente bianca e inuda," Vespucci to de' Medici, in Vespucci and Formisano *Lettere di Viaggio*, 16; English translation from Vespucci, *Letters from a New World*, 21. For the original spellings used by Vespucci, see Vespucci and Formisano, *Lettere di Viaggio*, 16–19; for the English translations of these place names, see Vespucci and Formisano, *Letters from a New World*, 21–25. The member of the expedition is believed to be Gaspar da India, whom Vasco da Gama brought back from India and who later accompanied Cabral; see Amado and Figueiredo, *Brazil 1500*, 37, n. 43, and 153, n. 44.

15. Quotation from Vespucci: "credo che sai la provincia che Tolomeo la chiama Gedrosica [sic]," in Vespucci and Formisano, *Lettere di Viaggio*, 18; English translation from Vespucci and Formisano, *Letters from a New World*, 23. In Ptolemy's *Geography*, the ninth map of Asia includes Gedrosia; see "Nona Asie tabula continet Ariam & Paropanisadas & Dragianam & Arachosia & *Gedrosiam*" (emphasis mine), in Ptolemy, *Cosmographia* (1486). Quotation from Vespucci: "Credete, Lorenzo, che quello che io ho scritto infino a qui è la verità, e se non si risconteranno le province e regni e nomi di città e d'isole colli scrittori antichi, è segnale che sono rimutati, come veggiamo nella nostra Europia, ché per maraviglia si sente uno nome antico," in Vespucci and Formisano, *Lettere di Viaggio*, 20; English translation in Vespucci and Formisano, *Letters from a New World*, 26.

16. Vespucci also mentiones "isola Trapobana" (Island of Taprobane) in his first letter to de' Medici; see Vespucci and Formisano, *Lettere di Viaggio*, 13. The island appears in the twelfth map of Aisa, "Duodecima Asie Tabula," in Ptolemy's *Geography*; see *Cosmographia* (1486). Quotation from Vespucci: "E io tengo speranza in questa mia navicazione rivedere e corere gran parte del sopradetto e discoprire molto più," in Vespucci and Formisano, *Lettere di Viaggio*, 19; English translation in Vespucci and Formisano, *Letters from a New World*, 24.

17. Metcalf, *Go-betweens*, 39–40; Dursteler, "Reverberations of the Voyages," 45–51. Letter of D. Cretico, Portugal, 27 June 1501, in Greenlee, *The Voyage of Pedro Alvares Cabral*, 119–23. Letter of Giovanni Francesco de Affaitadi to Domenico Pisani, Lisbon, 26 June 1501, in Greenlee, *The Voyage of Pedro Alvares Cabral*, 124–29. The Affaitadi were from Cremona, and they became active in Lisbon, first as investors in the sugar trade from Madeira in the late fifteenth century. Following Vasco da Gama's return, they also invested in the spice trade; see Denucé, *Inventaire des Affaitadi*, 17–20.

18. Quotation from Trevisan: "la qual sarà benissimo facta et copiosa, et particular de quanto paese è stato scoperto," in Angelo Trevisan to Domenico Malipiero, Granada, 21 August 1501, in Trevisan and Aricò, *Lettere sul Nuovo Mondo*, 28; on this letter, see also Harrisse, *Christophe Colomb*, 2:116–24. This correspondence is discussed by Dursteler, "Reverberations of the Voyages," 51–54. Quotation from Trevisan: "De carta de quel viazo [de Calicut] non è possibele haverne, che 'l Re ha messo pena la vita a chi le dà fora," Angelo Trevisan to Domenico Malipiero, Granada, 21 August 1501, in Trevisan and Aricò, *Lettere sul Nuovo Mondo*, 28; in English, see Greenlee, *The Voyages of Pedro Álvares Cabral*, 123.

Quotation from Trevisan: "Circa el desiderio ha la Magnificencia vostra de intender el viazo de Calicut, jo li ho scritto altre fiate che aspeto de zorno in zorno messier Cretico, qual me scrive haverne composto una opereta. Subito che 'l sia zonto, farò che la Magnificencia vosta ne haverà parte," Angelo Trevisan to Domenico Malipiero, Granada, September 1501(?), in Trevisan and Aricò, *Lettere sul Nuovo Mondo*, 34; in English, see Greenlee, *The Voyages of Pedro Álvares Cabral*, 123–24. Quotation from Trevisan: "Se venimo a Venetia vivi, vostra Magnificencia vedrà carte et fino a Calicut et de là più . . . Vi prometto che l'è venuto in ordene." Angelo Trevisan to Domenico Malipiero," Ecija, 3 December 1501, in Trevisan and Aricò, *Lettere sul Nuovo Mondo*, 41; in English, see Greenlee, *The Voyages of Pedro Álvares Cabral*, 124. On the various emissaries and ambassadors sent from Venice—Trevisan, Pisani, and Camerino "Il Cretico"—see Aricò, "Introduzione," in Trevisan and Aricò, *Lettere sul Nuovo Mondo*, 9–24, and Greenlee, *The Voyages of Pedro Álvares Cabral*, 114–18. On Pietro Pasqualigo specifically, see Weinstein, *Ambassador from Venice*. For the letter of Messer Cretico, see Berchet, *Fonti Italiane*, 83–86. The first edition of 1501 no longer exists, but it was subsequently published in 1507; see Amado and Figueiredo, *Brasil 1500*, 183.

19. This letter is dated Lisbon, 1502, and is available in Vespucci and Formisano, *Lettere di Viaggio*, 21–25, as "Nota d'una lettera venuto d'Amerigo Vespucci a Lorenzo di Piero Francesco de' Medici l'anno 1502 da Lisbona della loro tornata delle nuove terre mandato a cercare per la Maestà de re di Portogallo; e prima." The English translations are from Vespucci and Formisano, *Letters from a New World*, 29–35.

20. *Portolan Chart (King-Hamy)*, HM 00045, The Huntington Library, San Marino, CA, available online at http://ds.lib.berkeley.edu/HM00045_43 and http://dpg.lib .berkeley.edu/webdb/dsheh/heh_brf?CallNumber=HM+45. On this world chart, see Dutschke et al., *Guide to Medieval and Renaissance Manuscripts*, 1:103.

21. Duke Ercole d'Este had a network of informants who kept him abreast of the latest news, sending him *avissi* (newsy reports), often in the form of letters, from courts throughout Europe. Written in Italian, the *avissi* describe court events, such as weddings, births, and deaths, and they relate political news on possible preparations for war, the negotiation of treaties, and bits of secret information. More than six hundred *avissi* survive for a single year, 1497, in the Este family archives in Modena, indicating that Ercole was receiving a huge volume of correspondence from his many agents across Europe. Cantino does not seem to have been an official informant of the duke, for there is no record of his salary. Cantino may have hoped for a steady sinecure, or at least to be rewarded. For the correspondence, see Alberto Cantino in Chancelleria Ducale, Estero, Ambasciatori, Agenti e Corrispondenti Estensi, Fuori d'Italia—Spagna, Archivio di Stato di Modena (hereafter ASM). The letters are Cantino to d'Este, Cadiz, 19 July 1501, ASM 3988-74; Cantino to d'Este, Lisbon, 17 October, 1501, ASM 3988-74; this last letter is reprinted in Harisse, *Les Corte-Real*, 204–9; Cantino to d'Este, Lisbon, 30 January 1502, ASM 3988-74. The quotation from the letter of 1 October 1501: "liquali Io ho visti, tochi & contemplati;" in the same letter, the reference to those deserting the king's ships in Brazil were sailors in Cabral's armada headed to India.

22. On *Carta del Cantino*, see Cortesão and Mota, "Anónimo," in *Portugaliae Monumenta Cartographica*, 1:7; Fernandes, *O Planisfério de Cantino*; Gaspar, "From the Portolan Chart" and "Blunders, Errors, and Entanglements"; Milano, *La Carta del Cantino*; and Roukema, "Brazil in the Cantino Map"; and Leite, "O mais antigo mapa." The possibility

that Cantino somehow obtained the map commissioned by another is suggested by Gaspar, "From the Portolan Chart," 132. I argue that the anonymous chartmaker is likely Pedro Reinel; see Metcalf, "Who Cares Who Made the Map?" The price paid for the map is recorded in Cantino's 1502 letter; see Cantino to d'Este, Rome, 19 November, 1502, ASM 3988-74; this letter is transcribed in Harrisse, *Les Corte-Real*, 215–16. Cantino most likely rolled up the world chart and tied it closed and possibly inserted it into a long cylindrical case, as was done, for example, with the *Mappamondo Catalano estense,* which was stored in a *guaina grande de cuoio*—a leather scabbard or case in 1488, according to an Este inventory; see Sicilia and Bibi, *Alla scoperta del mondo,* 26. One example of a portolan chart case survives in the Biblioteca Nazionale Centrale in Rome; see the reproduction of the 84×12.5 cm case and discussion of it in Astengo, "Renaissance Chart Tradition," 183. On the discoveries of João da Nova, see "Nova, João da," in Albuquerque and Domingues, *Dicionário,* 2:805–6. The dedication on the back of the world chart reads: "Carta de navigar per le Isole nouam tr[ovate] in le parte de India: dono Alberto Cantino al S. Duca Hercole."

23. Cantino noted that he received twenty discounted ducats (*ducati vinte stretti*) from Francesco Catanio in Genoa. Quotation from Cantino's letter: "La Charta é di tal sorte / & spero che in tal manera piacera á V. Ex." in Cantino to d'Este, Rome, 19 November, 1502, ASM 3988-74; this letter is reprinted in Harisse, *Les Corte-Real*, 215–216. From this exchange, it would seem that the duke thought a high sum had been paid for the world chart, even if it was filled with the latest information from Lisbon. It also suggests that Cantino thought it a reasonable price, given its size, illustrations, and his efforts to obtain it.

24. Caverio, *Planisphère nautique / Opus Nicolay de Caverio ianuensis*, Département Cartes et plans, GE SH ARCH-1, Bibliothèque nationale de France http://gallica.bnf.fr /ark:/12148/btv1b550070757. Gregory McIntosh believes Caverio copied *Carta del Cantino* in Genoa, personal communication, July 2019. Similarly, Joaquim Alves Gaspar notes their remarkable similarities in "From the Portolan Chart." Alternatively, both might have shared a lost prototype; see Cortesão, *Cartografia portuguesa antiga,* 170. It also might have been copied later, as maps did circulate outside of the ducal library, at least during the tenure of Borso; see Milano and Battini, *Mapamundi Catalán Estense,* 14.

25. Rosselli and Contarini. *Mundu [sic] spericum . . . cognosces diligentia joani mathei Contareni, arte et ingenio francisci Roselli florentini 1506 notu,* Cartographic Items Maps C.2.cc.4. British Library, https://www.bl.uk/collection-items/first-known-printed-world -map-showing-america. On this map, see Heawood, "A Hitherto Unknown World Map."

26. Waldseemüller, *Universalis Cosmographia Secundum Ptholomaei Traditionem et Americi Vespucii Alioru[m]que Lustrationes,* 1507, Library of Congress, Washington, DC. https://www.loc.gov/resource/g3200.ct000725C/. The project, according to historian Christine Johnson, resulted from the "collective endeavours" of Lud, the editor and printer of the book; Ringmann, the writer of the book, who was assisted by Waldseemüller; and Waldseemüller, the maker of the wall map and globe. See "Renaissance German Cosmographers," 11; Waldseemüller and Hessler, *The Naming of America*; Karrow, *Mapmapers of the Sixteenth Century,* 568–72; and Meurer, "Cartography in the German Lands," 1204–7.

27. Several new versions of the Globe Gores have recently come to light, and one obtained by Christies was determined to be a forgery. Subsequently, the Globe Gores held at the Bavarian State Library was also determined to have been a fake; see Blanding,

"Why Experts Don't Believe"; Green, "The Discovery"; Jackson, "How a Rare Map"; and Bayerische StaatsBibliothek, "Bavarian State Library Confirms."

28. In one of the text boxes of the *Carta Marina* of 1516, Waldseemüller states that one thousand copies of a previous map had been printed; presumably, although not definitively, this refers to the 1507 wall map, the *Universalis Cosmographia*.

29. Buisseret, *Mapmakers' Quest*, 46. On the price of Rosselli's maps, see Carlton, *Worldly Consumers*, 66. On mapping as a "casually acquired skill" and "a common pastime," see Alpers, "Mapping Impulse," 60.

30. Carlton, *Wordly Consumers*, 22; Schulz, "Jacopo de' Barbari," 458. Although *View of Venice* has been seen as both a map and a view of the city, Schulz forcefully maintains that it is a work of art (441). On Leonardo's plan of Imola, and the increase in city plans, see Ballon and Friedman, "Portraying the City," 680, 682–83. On the blurring between maps and paintings, see Carlton, *Wordly Consumers*, 89; Buisseret, *Mapmakers' Quest*, 46; and Alpers, "Mapping Impulse," 54, 72.

31. Carlton, *Wordly Consumers*, 58; Woodward, "Italian Map Trade," 779–80; Van Duzer, "A Newly Discovered."

32. Ruysch, *Universalior cogniti orbis tabula ex recentibus confecta observationibus*, in Ptolemy, *Geographia*, 1508, John Carter Brown Library, Brown University, https://jcb .lunaimaging.com/luna/servlet/detail/JCBMAPS-1-1-1716-103410003:Universalior-cogniti -orbis-tabula-e?qvq=q:Ruysch&mi=0&trs=1; a copy in color is at the James Ford Bell Library at the University of Minnesota, https://www.lib.umn.edu/bell/maps/ptolemy3. On the Johannes Ruysch world map, see Nebenzahl, *Atlas of Columbus*, 48–51; McGuirk, "Ruysch World Map," 133; and Meurer, "Cartography in the German Lands," 1188–89, quotation on projection (1188).

Chapter 3 · Chartmakers

1. *Carta universal de Juan de la Cosa*, MNM-257, Museo Naval, Madrid & Biblioteca Virtual del Ministerio de Defensa, Madrid, http://bibliotecavirtualdefensa.es /BVMDefensa/i18n/consulta/registro.cmd?id=16822. When La Cosa made his declaration in the Acta de Pérez de Luna, 12 June 1494, that the coast of Cuba was part of the mainland and not an island, he declared himself to be a "maestro de hacer cartas"; see Navarrete, *Colección de los viages* (1825), 2:146. La Cosa's first voyage with Columbus was in 1492, on which he served as master of the *Santa Maria*. On the second voyage, also with Columbus, he served on the *Santa Clara*. His third voyage, with with Alonso de Ojeda, landed off the coastline of northern South America, south of the equator, and then traveled up the coast and into the Caribbean. On his fourth voyage, La Cosa sailed along the northern coast of South America and the Isthmus of Panama. La Cosa's fifth voyage to the Americas occurred in 1504; see Silió, *Carta de Juan de la Cosa*, 69. La Cosa never returned, as he was killed in February of 1510, along the mainland of northern Spanish America by an indigenous group resisting the Spanish slaving expedition in Turbaco, near Cartagena de Indias. According to Sandra Sáenz-Lopez Pérez, La Cosa did not paint the icon of the Virgin and Child himself but instead cut and pasted the image into the center of the compass rose and subsequently colored it; see Pérez, "La Carta de Juan de La Cosa," 13.

2. T. Campbell, "Census." On Mediterranean commerce extending to Northern Europe, see J. Cortesão, *Descobrimentos pré-colombinos*, 27–29, and Pamela O. Long, "World of Michael of Rhodes," in Long, McGee, and Stahl, *The Book of Michael of Rhodes*, 3:7 and map 1.1, p. 2.

3. The meaning of rhumb is given by the *Oxford English Dictionary* as "a line or course followed by a ship" and "an imaginary line on the earth's surface intersecting all meridians at the name angle and used as the standard method of plotting a ship's course on a chart" (*OED Online*, s.v. "rhumb"). On definitions of portolan charts and rutters, see Albuquerque and Domingues, *Dicionário*, s.v. "portulano" and s.v. "roteiro." Studies of the charts and rutters include Piero Falchetta, "Portolan of Michael of Rhodes," in Long, McGee, and Stahl, *The Book of Michael of Rhodes*, 3:193–210; T. Campbell, "Portolan Charts"; Mollat and La Roncière, *Sea Charts*, 11–35; T. Campbell, "Census"; Puchades i Bataller, *Les cartes portolanes*; Bagrow and Skelton, *History of Cartography*, 61–66; Astengo, "The Renaissance Chart Tradition," 174–237; Fernández-Armesto, "Maps and Exploration"; A. Cortesão and Mota, *Portugaliae Monumenta Cartographica*, 1:xxv–xxxv. "Living record" from T. Campbell, "Portolan Charts," 372–73. For a recent, highly critical review of much of the scholarship on medieval Mediterranean charts, see Nicolai, *The Enigma*, 11–20, 44–69, 231–35.

4. Edney, *Mapping an Empire*, 3. Edney is discussing Asia, but the point holds for the Atlantic. Edney's perceptive essay, "Knowledge and Cartography in the Early Atlantic," is key to the arguments presented in this chapter.

5. *Portolan Chart of the Mediterranean Sea*, ca. 1320–50, Geography and Map Division, G5672.M4P5 13—.P6. Library of Congress, Washington, DC, http://hdl.loc.gov/loc.gmd /g5672m.ct000821.

6. Puchades i Bataller, *Les cartes portolanes*, 497; Falchetta quoted by Edson, *The World Map*, 57; on Benincasa, see Astengo, "Renaissance Chart Tradition," 220, and Sheehan, "The Functions of Portolan Maps," 188–99.

7. Franciscus Becharius, *Portolan chart of the Mediterranean Sea, the North Atlantic Ocean, the Black Sea, and the northwestern African coast* [1403], Art Storage 1980 158, Beinecke Rare Book and Manuscript Library, Yale University, New Haven, CT, https://brbl-dl.library.yale .edu/vufind/Record/3521236. On this chartmaker, whose name is variously spelled as Beccario, Becario, Becaria, Becaro, Becaa, Becharius, see Skelton, "A Contract for World Maps"; and Unger, *Ships on Maps*, 45–46. On the description of him as a "mestre de cartes de navegar," see Document II (Barcelona, 26 May / 28 May 1400), in Skelton, "A Contract for World Maps," 111. For Grazioso Benincasa's *Portolan Chart of Europe*, see Additional MS, 31318A, British Library, London, https://www.bl.uk/collection-items/a-portolan-chart-of -europe-by-grazioso-benincasa-with-the-aegean. On Benincasa, who made charts and atlases in Ancona, Venice, Rome, and Genoa, see Caraci, "An Unknown Nautical Chart."

8. Skelton, "A Contract for World Maps," quotation from p. 108. According to Skelton, Jacome Riba was also known as Jafuda Cresques, the son of Abraham Cresques, the famous chartmaker believed to have made the *Atlas de cartes marines*. On the importance of this correspondence for understanding the illuminations on charts, as well as the intracies of chartmaking, see Unger, *Ships on Maps*, 45–46, and S. Davies, *Renaissance Ethnography*, 56.

9. "Sia bella e buona con le bandiere," in Michienzi and Vagnon, "Commissioning and Use of Charts," 24.

10. Mecia de Viladestes, *Carte marine de l'océan Atlantique Nord-Est, de la mer Méditerranée, de la mer Noire, de la mer Rouge, d'une partie de la mer Caspienne, du golfe*

Persique et de la mer Baltique, 1413, Cartes et plans, GE AA-566 (RES), Bibliothèque nationale de France, Paris, https://gallica.bnf.fr/ark:/12148/btv1b55007074s. The Atlantic portion of this chart is redrawn in figure 3.6. The legend for Gades reads: "Ilhes de gades. Les Iles de gades se oservien asi por Solonnio y por ysidorum" (The Islands of Gades as observed here from Solinio [Gaius Iulius Solinus] and Isidore [of Seville]). For a slightly alternate transcription, see A. Cortesão, "A Carta Náutica," 83. A legend written next to the whale hunt combines both fact and fantasy by stating that in the Ocean Sea such large fish are found that sailors think they are islands. For the legend on the whale hunt, see Van Duzer, *Sea Monsters*, 50–51. Van Duzer notes that the idea that whales could be mistaken for islands was a well-established myth (48). On this chart, see Pérez, "La pluralidad religiosa," and on the chartmaker, Hamy, "Mecia de Viladestes."

11. On the lack of coherence in the early Atlantic, where there is an enormous variety of microclimates, different disease environments, and thousands of languages, in contrast to the coherence of the Mediterranean, see Games, "Atlantic History," 742. Whereas sailing in the Mediterranean did not require sailing out of sight of land for long, that was not the case in the Atlantic; see Rodger, "Atlantic Seafaring," 75–77.

12. On early chartmaking in Portugal: A. Cortesão and Mota, *Portugaliae Monumenta Cartographica*, 1:xxx–xxxv, and Alegria et al., "Portuguese Cartography in the Renaissance," 977–80. The claim that Prince Henry brought "mestre Jacome" to Lisbon is made by Barros, *Asia, Primeira Decada*, bk. 1, chap. XVI, 1:61, and by D. Pereira, *Esmeraldo*, 58; it is also emphasized by Armando Cortesão in *História*, 2:94–98. Some Portuguese historians believe this was unlikely; see Marques, *Origem e desenvolvimento da cartografia*, 72–74, and Albuquerque, *Historia de la navegación*, 252–55. On Genoese influences in Portugal, see Marques, *Origem e desenvolvimento da cartografia*, 66–67.

13. Chaps. VIII, IX, LXXVI, and LXXVIII in Zurara, *Crónica*. The conversation between the prince and Eanes was on the eve of the rounding of Cape Bojador, which occurred in 1434; see *Crónica*, chap. IX, 87–91. Zurara's statement on charts ruling the seas is from chapter VIII: "que se regen todollos mares, per onde gentes podem navegar," *Crónica*, 48. On Zurara and the writing of the *Crónica*, see Linte, "As novas humanidades," 16–21. On Prince Henry, see Russell, *Prince Henry*. On fifteenth-century cartography in Portugal, see A. Cortesão and Mota, *Portugaliae Monumenta Cartographica*, 1:xxx–xxxi, and Marques, "The Dating of the Oldest Portuguese Charts." On the Canary Islands, see Abulafia, *Discovery of Mankind*, 33–101.

14. The legend is from *Il mappamondo di Fra Mauro*: "i qual la maiestà del Re de portogallo à mandato cum le suo caravele a çerchar e veder ad ochio, i qual dice haver circuito le spiaçe de garbin più de 2000 mia oltra el streto de çibelter. . . . E i diti hano fato nuove carte de quel navegar e hano posto nomi nuovi a fiumere, colfi, cavi, porti, di qual ne ho habuto copia." Transcription and English translation from Falchetta, *Fra Mauro's World Map*, 211. The *mappamundi* that Fra Mauro prepared for the Portuguese royal family was lost sometime after 1528. Marques maintains that the Fra Mauro *mappamundi* was commissioned by Prince Pedro of Portugal and that the map was kept in the castle of São Jorge, then the king's residence in Lisbon. There it may have been seen by Pero da Covilhã in 1487 and the German doctor and traveler Hieronimus Münzer in 1494; see "The Discoveries and the *Atlas Miller*," in Miró, *Atlas Miller*, 24–25, 33–34. On the disappearance of the map, see Ratti, "Lost Map of Fra Mauro," 83. According to the Visconde de Santarém, the map

was in the archives of the Alcobaça monastery until 1528, but sometime after that it disappeared and has not been seen since; see "Note sur le mappemonde de Fra Mauro," in Ministério da Cultura, *A historia da cartografia*, 47–48.

15. *Carta nautica: Francia, Spagna, Africa occidentali*, Biblioteca Estense Universitaria, Modena, https://www.gallerie-estensi.beniculturali.it/opere/carta-nautica-francia-spagna-africa-occidentali/. On this chart, see A. Cortesão and Mota, *Portugaliae Monumenta Cartographica* 1:3–4; Marques, *Origem e desenvolvimento*, 117–23; Marques, "Dating the Oldest Portuguese Charts," 90.

16. Aguiar, or Aguilar (as he is referenced by the Beinecke), *Portolan Chart of the Mediterranean Sea, the North Atlantic Ocean, the Black Sea, and the West African Coast as far south as Sierra Leone*, 1492, 30cea/1492, Beinecke Rare Book and Manuscript Library, Yale University, New Haven, CT, https://brbl-dl.library.yale.edu/vufind/Record/3433718. On this chart, see Guerreiro, *A carta náutica*. The legends on the chart: "Aquy esta a sera dos montes claros" and "aqueste mar nom he ruive de seu naturall mas a tera de baixo he."

17. The text by Nunes is titled *Tratado en defensam da carta de marear* and is cited here in the edition transcribed and edited by Francisco Maria Esteves Pereira as published in multiple numbers of the *Revista de Engenharia Militar* in 1911 and 1912; hereafter cited as Nunes and Pereira, *Tratado en defensam*; quotation from p. 355: "sendo ella [a carta de marear] ho melhor estromento que se podera achar: pera a nauegaçam: e descubrimento de terras." On Nunes, see J. Gaspar "From the Portolan Chart," 27–29; Almeida "Science during the Portuguese Maritime Discoveries," in Bleichmar et al., *Science in the Spanish and Portuguese Empires*, 78–92, and Randles, "From the Mediterranean Portulan," 116. The text by Chaves is titled "Quatri partitu en cosmographia pratica i por otro no[m]bre llamado Espeio de Navegantes" and is cited hereafter as Chaves, "Quatri partitu." The quotation is from Tratado Segundo, Capitulo 2 (hereafter II: cap. 2): "la carte de marear nos es asi como un espejoen la qual senos representa la imagen del mundo." On Chaves, see Lamb, *Quarti Partitu*. The text by Cortés is titled *Breve compendio de la esfera y del arte de navegar* and is hereafter cited as Cortés, *Breve compendio*. It later appeared in English in 1589 as *The Arte of Nauigation*.

18. Sandman, "Spanish Nautical Cartography," 1099–1101; Gaspar, "From the Portolan Chart," 27–29.

19. Vellum and parchment are fundamentally the same thing—a surface created from an animal skin that was used in the making of manuscripts; see Reed, *Nature and Making of Parchment*, 78. On the use of vellum to make medieval portolan charts, see Puchades i Bataller, *Les cartes portolanes*, 186–87, 472, and Astengo, "Renaissance Chart Tradition," 182.

20. Astengo, "Renaissance Chart Tradition," 188. *OED Online* s.vv. "Cinnabar," "gall," "lapis lazuli," "ochre," "ultramarine," "vellum," "verdigris," and "vermillion."

21. From Cortés: "Sobre el punto en que se cortan [dos lineas rectas] se a de hazer centro y sobre el dar um circulo oculto que casi occupe toda la carta," Cortés, *Breve compendio*, III: cap. 2. From Chaves: "e asi es acabado todo el lineamento de la dicha carta, que en comun sedize arumbado," Chaves, "Quatri partitu," II: cap. 2.

22. From the center point on Vallseca's chart, there extend thirty-two rhumb lines that reach the circumference of the invisible circle, and on that circumference sixteen sub-points appear, and from them extend further networks of rhumbs. Vallseca enhanced the center point with an eight-point compass rose, and one of the sub-points, in the far north, also has a half rose of thirty-two points. See Vallseca, *Carte marine de la mer*

Méditerranée et de la mer Noire, Gabriel Devallsecha la affeta en mallorcha an MCCCCXXXXVII, Département Cartes et plans, GE C-4607 (RES), https://gallica.bnf.fr /ark:/12148/btv1b53064893j, Bibliothèque nationale de France, Paris. On the Mediterranean Chart presumably made in Lisbon circa 1500 and known today as *Alte Welt*, the central point is decorated with a sixteen-point rose, and sub-roses appear on the invisible circumfrence that extends from the West African coast north of Scotland to the eastern edge of the Mediterranean. This chart was attributed to Pedro Reinel by A. Cortesão and Mota in *Portugaliae Monumenta Cartografica*, 1:23–24, but according to Kupčík this attribution was later withdrawn and the chart "has again been regarded as an anonymous Portuguese work"; see *Münchner Portolankarten: Kunstmann I–XIII, und zehn weitere Portolankarten / Munich Portolan Charts: Kunstmann I–XIII, and Ten Further Portolan Charts*, 110–114; the chart is nicely reproduced as addendum 4, and the original is in the collection of portolan charts, Cod. Icon 140, Bayerische Staatsbibliothek, Munich.

23. Reinel, *Portulan. Carte des côtes de l'Europe et de l'Afrique occidentales, dressée par Pedro Reinel vers 1485*, 2 Fi 1582 bis, Archives départementales de la Gironde, Bordeaux. On this chart, see Amaral, *Pedro Reinel me fez*.

24. Marques, "Dating the Oldest Portuguese Charts," 91–92. The reasons for adding these additional coastlines over the continent of Africa are unclear; see Amaral, *Pedro Reinel me fez*, 22–23.

25. On magnetic declination, see "Magnetic Declination," National Geophysical Data Center, National Oceanic and Atmospheric Administration, http://www.ngdc.noaa.gov /geomag/declination.shtml. For how magnetic declination was understood and how navigation was practiced in the sixteenth century, see Gaspar, "From the Portolan Chart," 11, 17–19. Patricia Seed and German Díaz analyze the 1468 Benincasa chart of the west coast of Africa and conclude that the errors in compass readings likely derived from incorrectly calibrated compasses, which in turn distorted the chart; see "Rewriting the History of Mapping." Also on the accumulation of errors, see Randles, "From the Mediterranean Portulan Chart," 115. For astronomical navigation, see Albuquerque, *Historia de la navegación*, and Luís de Albuquerque "A Navegação Astronómica," in Cortesão, *História da cartografia portuguesa*, 2:225–372. On the latitude chart, see Cortesão, *Cartografia portuguesa antiga*, 102, and Gaspar, "Myth of the Square Chart," 2.

26. Solar declination is the angular position of the sun at solar noon with respect to the plane of the equator, with North positive. The declination varies with the time of the year, from the Northern Hemisphere +23.75° at the summer solstice to -23.75° at the winter solstice. See Solar Declination, Solar Calculator Glossary, NOAA, Earth System Research Laboratory, Global Monitoring Division, https://www.esrl.noaa.gov/gmd/grad /solcalc/glossary.html#S. By the middle of the fifteenth century, there were three available sources for determining solar declination; see Proverbio, "Astronomical and Sailing Tables," 474. On latitude charts, see Gaspar, "From the Portolan Chart," 34, and "Myth of the Square Chart." The use of a latitude scale can be seen on the *Alte Welt* chart of the Mediterranean, circa 1500, where the degrees appear in the Atlantic just to the West of the Azores; see Kupčík, *Münchner Portolankarten / Munich Portolan Charts*, 110–14. A second example appears on the anonymous *Atlantische Küsten*, ca. 1505, roughly approximate to where the line of the Treaty of Tordesillas ran in the Atlantic, see Kupčík, *Münchner Portolankarten / Munich Portolan Charts*, 35–40.

27. From Cortés: "con una pluma delgada se tornara a señalar con tinta. Despues de la tinta enxuta: con una migaja de pan se limpiara todo lo del humo y quedara la costa asentada con tinta en la carta." Cortés, *Breve compendio,* III: cap. 2. On copying coastlines from chart to chart, see Gaspar, "From the Portolan Chart," 29. Cortés's instructions are discussed by Astengo, "Renaissance Chart Tradition," 185, and Puchades i Bataller, *Les cartes portolanes,* 471–85; on the suggestion that silk was used, see 214, 479. *OED Online,* s.v. "pounce." See also T. Campbell, "Portolan Charts," 391, and T. Campbell, "Maphistory/History of Cartography, Workshops," http://www.maphistory.info/PortolanAttributions.html.

28. From Chaves: "deven se tener para esto muy buenos compases y los tales puntos señalarlos primero con plomo y despues con tinta trayendo la pluma muy delgada y la señal que hiziere muy subtil"; Chaves, "Quatri partitu," II: cap. 2.

29. On entering the toponyms on the coastline, Chaves, "Quatri partitu," II: cap. 2; Cortés, *Breve compendio,* III: cap. 2; Puchades i Bataller, *Les cartes portulanes,* 471–72.

30. D. Pereira, *Esmeraldo de Situ Orbis,* 42; 44.

31. Chaves, "Quatri partitu," II: cap. 2; Cortés, *Breve compendio,* III, cap. 2. According to Cortés's acount, the steps taken by chartmakers are similar to those described by Chaves, although not necessarily in the same order. Cortés indicates how the sub-roses are to be drawn, provides a detailed description of how the coastlines are to be inserted, and suggests the kinds of illustrations to be included, such as cityscapes, ships, and flags. He gives as the last steps the creation of scales and their placement on the chart, Cortés, III: cap. 2. Puchades i Bataller suggests that the chartmaker would plan to do all of the inking in the same color at once; see *Les cartes portulanes,* 474.

32. Titles of Pedro Reinel and his son Jorge Reinel from Viterbo, *Trabalhos nauticos,* 262, 158; for Lopo Homen, 29. Chaves and Cortés each provide detailed instructions for making a compass to be taken to sea: Chaves, "De la fabrica del aguja del marear," in "Quatri partitu," II: cap. 1. Cortés, "de la fabrica de la aguja—o brujola [brújula] de navegar," in *Breve compendio,* parte III, cap. 4.

33. Astengo, "Renaissance Chart Tradition," 192; Cortesão, *Cartografia portuguesa antiga,* 159–60; Chaves, "Quatri partitu," II: cap. 2. Decorative portolans began to appear early on, and because most of the portolan charts that survive have a decorative quality, Quinn argues that they were preserved in part because of their artistic qualities; see "Artists and Illustrators," 244.

34. According to Duarte Pacheco Pereira, the name given to Sierra Leone had nothing to do with lions; rather, it referred to the wild nature of its coastline, as described by Pedro de Sintra, who sailed along the coast in 1460; see *Esmeraldo de Situ Orbis,* 57.

Chapter 4 · From Manuscript to Print

1. Buisseret, *Mapmakers' Quest,* 46.

2. Wood, Fels, and Krygier, *Rethinking the Power of Maps,* 23 (emphasis in original) and 22–27; Harvey, *Maps in Tudor England,* 7; Karrow, "Centers of Map Publishing," 621; Woodward, "Techniques of Map Engraving," 609. Similarly, Juergen Schulz writes that the "output of printed maps leapt from a handful to hundreds per year, and by the end of the sixteenth century the interest in maps had spread to embrace a general public of educated men"; see "Maps as Metaphors," 97. S. Davies, *Renaissance Ethnography,* 4; 62.

3. Shirley, *The Mapping of the World*, 2–10; Ptolemy, *Cosmographia*, 1482. On the printed editions of Ptolemy, and their influence, see Weiss, "*Geography* in Print"; and Zur Shalev, "Main Themes," in Shalev and Burnett, *Ptolemy's Geography*, 1–14. On copperplate engraving, see Verner, "Copperplate Printing, 51–52"; for the woodcut process, see Woodward, "The Woodcut Technique," 49.

4. Shirley, *The Mapping of the World*, 2–10, quotation from p. 9. "Insculptum est per Johannes Schnitzer de Armszheim" is placed at the top of the map; see the example at the James Ford Bell Library, University of Minnesota http://gallery.lib.umn.edu/archive /original/762c5731e77392a41396f74adb9dcc20.jpg.

5. On the challenges faced by printers, see Pettegree, *Book in the Renaissance*, 21–42, quotation from p. 42. Schedel, *Liber Chronicarum* and *Das Buch der Croniken und Geschichten*.

6. An example of a broadside map is that by Hanns Rüst, ca. 1480, titled *Das ist die mapa mu[n]di un[d] alle land un[d] kungk reich wie sie ligend in der ga[n]sze welt*, Incunable Collection, PML 19921, Morgan Library and Museum, New York, https://www.themorgan .org/incunables/145336. On this map, see Edson, *The World Map*, 173–75; Shirley, *The Mapping of the World*, 7–8; and Campbell and Destombes, *The Earliest Printed Maps*, 79–84.

7. Lud, a canon and secretary to René, was the editor and printer of the book, while Ringmann is thought to have authored the text (some of it with Waldseemüller), and Waldseemüller is recognized as the maker of the wall map and globe. St. Dié is in the Vosges Mountains, near Strasbourg. On the importance of this collaboration, see the preface and introduction to Waldseemüller et al., *Cosmographiae Introductio*; C. Johnson, "Renaissance German Cosmographers"; Waldseemüller and Hessler, *The Naming of America*, 39–66; John W. Hessler, "An island," in Hessler and Van Duzer, *Seeing the World Anew*; Karrow, *Mapmapers of the Sixteenth Century*, 568–72, Meurer, "Cartography in the German Lands," 1204–7, and R. A. Skelton, "Biographical Note," in Ptolemy and Skelton, *Geographia: Strassburg, 1513*, v–xiv.

8. "Comparavi enim mihi ante paucos dies pro aere modico sphaeram orbis pulchram in quantitate parva, nuper Argentinae impressam, simul et in magna dispositione globum terrae in plano expansum, cum insulis et regionibus noviter ab Americo Vesputio Hispano inventis," in Trithemii, *Epistolae Familiares*, xli, in Trithemius, *Opera historica*, 553. Trithemius's letter is cited by Waldseemüller, Fischer, and Wieser, *Älteste Karte*, 15–16, and by Herbermann, "The Waldseemüller Map," 328. Trithemius, *In Praise of Scribes*; Grafton, "How Revolutionary Was the Print Revolution?," 84.

9. Lorenz Fries, who printed a smaller version of Waldseemüller's *Carta Marina*, provided instructions for mounting a wall map; see Waldseemüller and Hessler, *Naming of America*, 11. The globe gores at the James Ford Bell Library at the University of Minnesota, Minneapolis, measure 18 × 35 cm, which would make a globe of approximately 35 cm (13.7 inches) in circumference, about the size of a bocce ball. See Hamlin and Ragnow, "Martin Waldseemüller & the Map." In Green, "The Discovery," an image of a reconstructed globe can be seen, albeit from a copy of the gores subsequently determined to be a forgery. On the forgery, see Blanding, "Why Experts Don't Believe." The popularity of wall maps, painted in Renaissance architectural spaces, especially the palaces of Italian princes, is beautifully presented by Francesca Fiorani in *The Marvel of Maps*.

10. On the quadrant, the instrument held by Ptolemy, see Evans, *The History and Practice of Ancient Astronomy*, 206. Renaissance mapmakers could use Ptolemy as a nonideological

framework into which they could insert new information, understanding that, as Ptolemy himself argued, knowledge needed to be updated constantly; see Grafton, Shelford, and Siraisi, *New Worlds, Ancient Texts*, 48–54; and C. Johnson "Renaissance German Cosmographers," 14–15. Johnson suggests that "Vespucci was not displayed because of his unique achievements but rather because he represented a type: the voyager in the Renaissance romance of discovery" (26).

11. The tropics appear on Waldseemüller's globe at 30° N and S respectively, but their placement on modern maps is 23° 26′.

12. The original title of the book written by Ringmann and Waldseemüller and printed by Lud reads in its totality: *Cosmographiae Introductio cum quibusdam geometriae ac astronomiae principiis ad eam rem necessariis: insuper quattuor Americi Vespucij nauigationes: Uniuersalis Cosmographiae descriptio tam in solido quam plano eis etiam insertis quae Ptholomeao ignota a nuperis reperta sunt.* The English translation is: *Introduction to Cosmography with Certain Necessary Principles of Geometry and Astronomy to which are Added the Four Voyages of Amerigo Vespucci, a Representation of the Entire World, both in the Solid and Projected on the Plane, Including also Lands which were Unknown to Ptolemy, and have been Recently Discovered.* See Waldseemüller et al., *Cosmographiae Introductio* (1907), I; 31. The poem in Latin by Ringmann addresses Maximilian as a great king, "sacred throughout the vast world," and dedicates the world map to him; see the translation by Fischer and Wieser in Waldseemüller et al., *Cosmographiae Introductio* (1907), 32. Lehmann, in "The Depiction of America," argues that the dedication to Maximilian was linked to the desire to protect the interests of German merchants in the spice trade. The Latin text in which Waldseemüller describes the map and the globe as for scholars is: "totius orbis typum tam in solido quam plano (velut praeviam quandam ysagogen) pro communi studiosorum utilitate paraverim"; as translated by Fischer and Wieser: "I have prepared for the general use of scholars a map of the whole world—like an introduction, so to speak—both in the solid and projected on the plane," in Waldseemüller et al., *Cosmographiae Introductio* (1907), 33–34.

13. The written map, Jacob argues, "constitutes an important medium of cultural transmission: such a map is a mediation between the public and a preexisting or contemporary library that it both quotes and sums up"; *The Sovereign Map*, 191–92.

14. Quotation from Jacob, *The Sovereign Map*, 192. Trithemius: "et regionibus noviter ab Americo Vesputio Hispano inventis"—"and new regions recently discovered by Amerigo Vespucci of Spain," in Trithemius, *Epistolae Familiares*, xli, in *Opera historica*, 553.

15. Jacob, *The Sovereign Map*, 240; 242; see also Jacob's discussion of map legends on the margins, 263–66, which often integrate the map into historical time, as do Waldseemüller's. The Latin from the legend in top left box: "Est enim terra per Columbum regis Castiliae capitaneum atque Americum Vesputium . . . inventa . . . atque inter tropicos iaceat nihilo tamen minus ad undeviginti ferme gradus ultra caprcornum ad polum Antarcticum," as translated by John Hessler, in Waldseemüller and Hessler, *Naming of America,* 13. Latin from legend in top right box: "veterum inventa ponere / et ea quae a neotericis interim reperta sunt . . . coniungere placuit," as translated by John Hessler, in Waldseemüller and Hessler, *Naming of America,* 30. Latin from legend in lower-right box: "Non tamen parum ipsis eisdem incognita manserunt, sicut est in occasu Americae: ab eius nominis inventore dicta, quae orbis quarta pars putanda est," and "Id autem unum rogamus ut rudes et cosmographiac ignari hacc non statim damnent anteaquam didicer-

int chariora [clariora] ipsis haud dubie post cum intellexerint futura," as translated by John Hessler, in Waldseemüller and Hessler, *Naming of America*, 34.

16. The authors of the *Cosmographiae Introductio* are thought to be Ringmann and Waldseemüller, and the book consists of two distinct parts. Part 1 delivers a basic course on geometry, the five climatic zones of the earth, and the nature of the winds. These chapters establish Ptolemy's "system," which is needed to understand the new information that will be introduced in chapter 9. In this last chapter, the humanists present the information on the fourth part of the world; see Waldseemüller et al., *Cosmographiae Introductio* (1907), and Waldseemüller and Hessler, *The Naming of America*. The selection of the name America appears in chapters 7 and 9. Part 2 of the *Cosmographiae Introductio* presents a translation into Latin of Amerigo Vespucci's second published letter, known as the *Lettera*, the *Four Voyages*, or the *Soderini Letter*.

17. The Gonfalonier of Justice presided over the most important institution of government in Florence; see "Gonfaloni, Gonfalonieri di Giustizia," in Schweitzer and Wedeck, *Dictionary of the Renaissance*, 277. Vespucci's letter is altered so that Vespucci addresses their own patron, Duke René. For the text of the letter in Latin, see Waldseemüller et al., *Cosmographiae Introductio*, pt. 2; in Italian, see Vespucci and Formisano, *Lettere di Viaggio*, 37–66; for English, Waldseemüller et al., *Cosmographiae Introductio*, 83–151, and Vespucci and Formisano, *Letters from a New World*, 57–97.

18. For Las Casas's critique of Vespucci's four voyages, see *Historia de las Indias* 2:268–269, 271–74, 354, 390–418, 426. The scholarly unease with Vespucci is well reflected in Fernández-Armesto, *Amerigo*. Fernández-Armesto argues that Vespucci made only two voyages; see *Amerigo*, 62–93. Also on Vespucci, see Arciniegas, *Amerigo and the New World*; Gerbi, *Nature*, 35–49; Formisano, introduction to Vespucci and Formisano, *Letters from a New World*, xix–xl; and Markham, introduction to *Letters of Amerigo Vespucci*, i–xliv. Vespucci's unpublished or "familiar" letters, addressed to Lorenzo di Pierfrancesco de' Medici, are considered by scholars to be more reliable than the published letters; see Formisano, introduction to Vespucci and Formisano, *Letters from a New World*, xxvi–xxvii. Also see my chapter 6, where these texts, as well as their descriptions of cannibalism, are addressed.

19. Waldseemüller et al., *Cosmographiae Introductio* (1907), xxix–xxx.

20. Latin text for first quotation: "Et quarta orbis pars (quam quia Americus invenit Amerigen / quasi Americi terram / sive Americam nuncupare licet)" as translated by Fischer, Wieser, and Herbermann, in Waldseemüller et al., *Cosmographiae Introductio* (1907), 63. Latin text for second quotation: "Quam non video cur quis iure vetet ab Americo inventore sagacis ingenii viro Amerigen quasi Americi terram / sive Americam dicendam cum et Europa et Asia a mulieribus sua sortitae sint nomina," as translated by John Hessler in Waldseemüller and Hessler, *Naming of America*, 101.

21. Latin text from *Cosmographiae Introductio*: "terra iam quadripartita cognoscitur: et sunt tres primae partes continentes; quarta est insula, cum omni quaque mari circumdata conspiciatur," as translated by John Hessler in Waldseemüller and Hessler, *Naming of America*, 101. Latin from legend in lower-left box: "Terrarum insularumque variarum generalis descriptio etiam quarum vetusti non meminerunt auctores nuper ab anno domini 1497 usque ad 1504," as translated by John Hessler in Waldseemüller and Hessler, *Naming of America*, 17.

22. Latin for first quotation from *Cosmographiae Introductio* "In solido vero quod plano additur descriptionem Americi subsequentem sectati fuerimus," as translated by John Hessler

in Waldseemüller and Hessler, *Naming of America,* 107. Latin for second quotation from *Cosmographiae Introductio* "Fuit igitur necesse (quod ipse sibi etiam faciundum [faciendum] ait) ad novas temporis nostri traditiones magis intendere. Et ita quidem temporavimus rem, ut in plano circa novas terras et alia quaepiam Ptholomaeum," as translated by John Hessler in Waldseemüller and Hessler, *Naming of America,* 106–7.

23. The three humanists envisioned an edition of Ptolemy's text, which they would name *Geographia* after its original Greek title; it would correct errors in the Latin text, augment the tables with Greek names, and add new maps. Waldseemüller not only would design the usual maps that were then standard in an edition of Ptolemy's *Cosmographia/ Geographia* but would create *tabulae novae* (new maps) to supplement them. Waldseemüller was working on both the large wall map and the globe at the same time, as can be seen in his letter to the printer Johann Amerbach of Basel in 1507, where he stated that he was soon to print an edition of the *Geography* with new maps. No edition of the *Geographia* was attributed to the humanists of St. Dié at the time, but scholars believe the *Geography* printed in Strasbourg in 1519 was the edition researched by Ringmann and Waldseemüller; see Skelton, "Biographical Note," in Ptolemy and Skelton, *Geographia: Strassburg, 1513,* v–vi. The edition is *Claudii Ptolemei viri Alexandrini mathematice discipline philosophi doctissimi Geographie opus nouissima traductione e Grecorum archetypis castigatissime pressum*: see the copy at the Library of Congress, Washington, DC, https://www.loc.gov/item/48041351/. On Waldseemüller's sources, see Skelton, "Bibliographical Note," in Ptolemy and Skelton, *Geographia: Strassburg, 1513,* xvii; Van Duzer, "Multispectral Imaging"; Waldseemüller, Fischer, and Wieser, *Die Älteste Karte,* 27–29; Herbermann, "The Waldseemüller Map of 1507," 335–37; Fisher and Weiser, introduction to Waldseemüller et al., *Cosmographiae Introductio* (1907), 21; McIntosh, *The Vesconte Maggiolo World Map*; and Hessler, "An island," in Hessler and Van Duzer, *Seeing the World Anew,* 8–10.

24. Robinson, "Mapmaking and Map Printing," 3–4.

25. Barbari, *View of Venice,* Cleveland Museum of Art, http://www.clevelandart.org /art/1949.565.1. On this image, see Schulz, "Jacopo de' Barbari"; Howard, "Venice as a Dolphin," 101–3; Landau and Parshall, *Renaissance Print,* 43–46; and Hind, *Some Early Italian Engravings,* 285. Although de' Barbari's *View of Venice* has been characterized as a woodcut map, Schulz forcefully maintains that it is a work of art, in "Jacopo de' Barbari," 441. Following the completion of the map, Barbari became a link between northern Italy and Germany when he traveled to Nuremberg, where he entered the service of the Holy Roman Emperor.

26. Landau and Parshall stress the different specializations of the designer and the woodcutter; see *Renaissance Print,* 33. On the large wall map being printed in Strasbourg, see Waldseemüller, Fischer, and Wieser, *Die Älteste Karte,* 17; and Shirley, *The Mapping of the World,* 28. On the uncertainty on where the map was printed, as well as Hessler's assertion that it was printed by Johannes Schott in Strasbourg, see Hessler, "An island," in Hessler and Van Duzer, *Seeing the World Anew,* 11–12.

27. Berggren, Jones, and Ptolemy, *Ptolemy's Geography,* 86–93; Hessler, "What Can Waldseemüller's Projection Tell Us?" and "Infinite Geometries."

28. Berggren, Jones, and Ptolemy, *Ptolemy's Geography,* 94–95.

29. Grenacher, "The Woodcut Map," 31; quotation from Landau and Parshall, *Renaissance Print,* 242; Robinson, "Mapmaking and Map Printing," 17.

30. Harris, "Waldseemüller World Map," 36 and fig. 3. Cletes were used for the large woodblocks for Jacopo de' Barbari's huge woodcut map of Venice, which have survived and are on display in the Museo Correr of Venice. Woodward, "Techniques of Map Engraving," 593, fig. 22.3. Dürer's larger woodblocks were also held together with metal cletes; see Landau and Parshall, *Renaissance Print*, 22.

31. On the preparation of the surface and the transfer process, see Landau and Parshall, *Renaissance Print*, 22–23; Stewart, *Before Bruegel*, 242; Woodward, "The Woodcut Technique," 43.

32. Grenacher, "The Woodcut Map," 31; Skelton's comment in "Bibliographical Note," in Ptolemy and Skelton, *Geographia: Strassburg, 1513*, xviii.

33. The winds are discussed in chapter 8 of *Cosmographiae Introductio*; see also Waldseemüller and Hessler, *Naming of America*, 95–99 and 114, n. 53, on the German poet. The quotation on the north wind is from Waldseemüller et al., *Cosmographiae Introductio* (1907), 66. For windheads on *mappaemundi*, see the windheads on Nicolò Germano's *Planisfero* in the *Cosmographia* (1466) owned by Borso d'Este (plate 3). Windheads also circle the world maps printed in the editions of Ptolemy's *Cosmographia*, such as in the Bologna edition (1477), the Rome editions (1478 and 1480), and the Ulm edition (1482). Windheads circle the *mappamundi* in the *Liber Chronicarum* (1493) and the *Margarita Philosophica* by Gregor Reisch, 1503, as well as on world map by Giovanni Contarini and Francesco Rosselli, 1506; see Shirley, *The Mapping of the World*, 4–25.

34. See Hess and Eser, *The Early Dürer*, 434–53; and Landau and Parshall, *Renaissance Print*, 42. Before the *Apocalypse* series, Panofsky describes the woodcutter's technique as consisting of two separate types of lines: thicker lines that served to define shapes and forms, and finer lines (cross-hatchings) to create light, shade, and texture. Rarely did these work together. Panofsky describes Dürer's approach as resting on the principles of "concentration and dramatization" because Dürer sought make the visions of the book of Revelation "convincing without the aid of conventional signs or inscriptions." *Albrecht Dürer*, 1:47–55; quotations from 47, 55. A simple comparison between Dürer's depiction of the sky with its clouds, as well as the windheads, in several *Apocalypse* prints and the border of Waldseemüller's map shows many similarities. In *St. John before God and the Elders*, Dürer positions God and John in an undulating and curling cloudbank. In *The Four Horsemen of the Apocalypse*, the sky is cut with thin, closely spaced, black lines to suggest darkness, while curving lines with short cut lines create the contrasting puffy white clouds. In *Revelation of St. John: Four Angels Holding the Winds*, four windheads appear in the four corners of the sky. Three of the windheads are three-dimensional; they burst forth from the sky blowing wind from their thick, full cheeks, while a fourth windhead is shown in profile. For the prints from this series, see Dürer and Panofsky, *Dürer's Apocalypse*.

35. That Dürer might have worked on the *Carta Marina* was suggested by Joseph Fischer; see Waldseemüller, Fischer, and Wieser, *Die Älteste Karte*, 19. On the possible participation of Dürer on the 1507 map, see Hessler, "An island," in Hessler and Van Duzer, *Seeing the World Anew*, 11. On Dürer's whereabouts, see Howard, "Venice as a Dolphin," 101–3; Landau and Parshall, *Renaissance Print*, 43–46; Hind, *Some Early Italian Engravings*, 285; and Dürer and Conway, *Writings of Albrecht Dürer*, 46–60. Baldung was born to a prominent family in Strasbourg in 1509 and was known to have a close association with Dürer (it is believed that Baldung was part of Dürer's Nuremberg workshop from 1503 to

1507) before Baldung returned to Strasbourg; see Sullivan, "The Witches of Dürer," 364–65. Wechtlin's first woodcuts appeared as illustrations in books printed by Grüninger in 1502, and three years later he was known to have been in the employ of Duke René of Lorraine for one year: 1505. He became a citizen of Strasbourg in 1514. See Bartrum, *German Renaissance Prints*, 64–80.

36. Jacob, *The Sovereign Map*, 107–13; quotations from 113.

37. On the creation of woodcut maps in early sixteenth-century Germany, and the insertion of hand-cut text, see Landau and Parshall, *Renaissance Print*, 23; 240–44; Hind, *Introduction to a History of Woodcut*, 1:7–8; Grenacher, "The Woodcut Map"; and Woodward, "Manuscript, Engraved, and Typographic Traditions," in Woodward, *Art and Cartography*, 183–84.

38. Harris, "Waldseemüller World Map," 30, 37; Carter, *A View of Early Typography*, 43–49, 88–90.

39. Pettegree, *Book in the Renaissance*, 26–28.

40. For Waldseemüller's discussion of color in *Cosmographiae Introductio*, see Waldseemüller and Hessler, *Naming of America*, 110. There is color on the surviving copy, a grid of red lines in brazilwood ink from Waldseemüller's time. However, these were not part of the original map and were possibly added by Schöner; see Wanser, "Treatment and Preparation."

41. Harris, "Waldseemüller World Map," 42–47, quotations from pp. 42, 47. Harris speculates that the typesetting for the map began in St. Dié but was finished in Strasbourg. The text set in roman type, she argues, contained few errors in the book or on the map. Therefore, one can conclude either that the printer knew Latin or that the text had been carefully proofread by those who knew Latin, presumably Waldseemüller, Ringmann, or Lud. This suggests that the roman type was set in St. Dié. Moreover, Johann Grüninger, an active and important printer in Strasbourg, did not own the roman type. Grüninger did, however, own the Gothic type visible on the map, and Grüninger did not read Latin; see Harris, "Waldseemüller World Map," 42–47; on Grüninger, see Johnson, *Carta marina*, 19–24; on his inability to read Latin, 18.

42. On the likelihood that the map was printed in Strasbourg, see Waldseemüller, Fischer, and Wieser, *Die Älteste Karte*, 17; and Shirley, *The Mapping of the World*, 28. For Schott, see Hessler, Karnes, and Wanser, "Who Printed Waldseemüller?," and Hessler, "An island," in Hessler and Van Duzer, *Seeing the World Anew*, 11–12. As Harris notes, Schott owned the roman typeface that was used to print the *Cosmographiae Introductio* and which appears in the title and five legends on the 1507 map; "Waldseemüller World Map," 42–47.

43. For Waldseemüller's letter, see Skelton, "Biographical Note," in Ptolemy and Skelton, *Geographia: Strassburg, 1513*, xii. Harris and others describe the watermark for all twelve sheets of the *Universalis Cosmographia* as a crown. Moreover, the crowns are just a touch different so that it is possible to distinguish two versions of the same watermark. This suggests, Harris and others argue, that the paper came from two molds—a pair—owned by the same papermaker. Hessler, Karnes, and Wanser, "Who Printed Waldseemüller?," and Harris, "Waldseemüller World Map," 32.

44. Schöner included a set of the gores for his celestial globe, a star chart now attributed to Albrecht Dürer, the twelve sheets from Waldseemüller's 1507 *Universalis Cosmographia*, and the twelve sheets of Waldseemüller's later world map printed in 1516, the *Carta Marina*. See Hessler, *Renaissance Globemaker's Toolbox*, 21; and Hessler, "An island," in

Hessler and Van Duzer, *Seeing the World Anew*, 3–6. Hessler notes the importance of the bookplate that Schöner pasted into the *Sammelband*, as translated by Hessler: "Schöner gives this to you posterity; as long as it exists there is a monument to his spirit"; Hessler, *Renaissance Globemaker's Toolbox*, 35. In other writings, Schöner was concerned that scientific knowledge could be lost or repressed; see Hessler, *Renaissance Globemaker's Toolbox*, 35–40, and Hessler, "An island," in Hessler and Van Duzer, *Seeing the World Anew*, 3–6.

Chapter 5 · Parrots and Trees

1. For the definition of codes and discussion of the ten cartographic codes, see Wood with Fels, *Power of Maps*, 54 and 124; and 117–24 (the quotation from p. 122). Harley, "Introduction: Texts and Contexts," 4–5, notes that "like all other texts, maps use signs to represent the world" and that when signs "become fixed in a map genre, we define them as conventional signs." Jacob's discussion of graphics, geometry, and figuration is especially interesting for historians; see *The Sovereign Map*, 103–87 (quotation is from p. 185).

2. Quotations from Columbus's Diario: "Puestos en la tierra vieron árboles muy verdes," Jueves 11 de Octubre; "junto con la dicha isleta están huertas de árboles las más hermosas que yo vi," Lunes 15 de Octubre; "que veo mil maneras de árboles," Lunes 22 de Octubre; "Dice el Almirante que nunca tan hermosa coso vido, lleno de árboles . . . fermosos y verdes, y diversos de los nuestros, con flores y con su fruto, cada uno de su manera," Domingo 28 de Octubre, in Colon, *Relaciones*, 23, 28, 43, 46; English translation from Columbus, *Diario*, 62–63, 77, 117. On La Cosas's geographic information, see Silió, *La Carta de Juan de La Cosa*, 105.

3. Parrots arrived in Lisbon as early as 1501; see "1ª Carta de Bartolomeu Marchionni," in Amado and Figueiredo, *Brasil 1500*, 188, and "Carta de João Matteo Crético," in Amado and Figueiredo, *Brasil 1500*, 177–78.

4. Boehrer, *Parrot Culture*, ix–57.

5. Columbus, *Diario*, 65; Gerbi, *Nature in the New World*, 15. Las Casas, *Historia de las Indias*, provides accounts on Columbus's first voyage: "Passaban tantas manadas de papagayos que cubrian el sol" (1:313); on parrots given as gifts (1:372); and on his second voyage on Santa María island: "y papagayos y otras muchas aves" (2:57); on reaching Dominica island "ven infinitos papagayos verdes" (2:5); on Guadalupe Island "Alli hallaron los primeros papagayos que llamaban guacamayos, tan grandes como gallos, de muchos colores" (2:6, 126, 240); descriptions of different parrots on the island of Trinidad, Gracia, Española, Cuba, Jamaica, and San Juan, where "parece que son en algo diferentes los de cada isla," and on the mainland, where "hay una especie de papagayos que creo que no hay en otra parte, muy grandes, poco ménos que gallos, todos colorados con algunas plumas, en las alas, azules y algunas prietas" (2:235–36); "inmensidad de papagayos verdes" in Cuba (5:298); small birds called *xaxabis* on Española (5:409); parrots given as gifts (2:438, 458; lists of parrots as food like fish or rabbits (2:438); how parrots are captured (3: 471); and how more than ten thousand parrots were eaten over fifteen days in Cuba (4:30).

6. Las Casas, *Historia de las Indias*: "Hizoles [Sus Altezas] un buen presente de oro . . . y muchos papagayos" (2:129); description of parrots brought back to Spain by Anghiera, *De Orbe Novo* (1516), Oceaneae Decadis, chap. 1, and Crediti Continentis, chap. 9; for English see Anghiera, *De Orbe Novo* (1912), 1:65, 265; Oviedo's account in *Historia general e natural*, 1:29.

7. "Carta de Pero Vaz de Caminha," in Amado and Figueiredo, *Brasil 1500,* 104; "Relação do Português Anônimo: 1500," in Amado and Figueiredo, *Brasil 1500,* 135–36. Other letters now lost were clearly written, as can be deduced from the third surviving document, in which Mestre João, the king's physician, states that others were writing at length to the king about everything that took place in Brazil; see "Carta de Mestre João," in Amado and Figueiredo, *Brasil 1500,* 123. Quotation on the parrots: "Há muitas aves de várias espécias, especialmente papagaios de muitas cores, entre os quais alguns grandes como galinha, e outros pássaros mui bonitos," "Relação do Português Anônimo: 1500," in Amado and Figueiredo, *Brasil 1500,* 136.

8. "Letters sent by Bartolomeo Marchioni to Florence," in Greenlee, *Voyage of Pedro Alvares Cabral,* 147–49 (quotation from p. 148). Marchionni was a wealthy Florentine in Lisbon who invested in the early Atlantic slave trade from Africa as well as in the India trade and Cabral's voyage of 1500; see Guidi-Bruscoli, "Bartolomeo Marchionni"; Thomas, *The Slave Trade,* 84–86, and Greenlee, *Voyage of Pedro Alvares Cabral,* 145–47. The letter by Marchionni is dated 27 June 1501 and exists as a manuscript copy of a lost original; see Greenlee, *Voyage of Pedro Alvares Cabral,* 147, 203. "Letter of Giovanni Matteo Cretico," as transcribed and translated by Greenlee, *Voyage of Pedro Alvares Cabral,* 119–23, quotation from p. 120. The letter is dated 27 June 1501.

9. Dispatch of Pietro Pasqualigo, 18 October 1501, to the Senate of Venice, as extracted from the *Diarii* of Sanuto, as transcribed and translated by Harrisse, *Les Corte-Real,* 209–11. For a partial Portuguese translation, see "Carta de Pedro Pasqualigo," in Amado and Figueiredo, *Brasil 1500,* 254. "Carta de Pero Vaz de Caminha," in Amado and Figueiredo, *Brasil 1500,* 77, and also n. 36. Caminha slightly alters the name of the newly discovered land as he concludes his letter, writing: "Beijo as mãos de Vossa Alteza. Deste Porto Seguro da vossa Ilha da Vera Cruz, hoje sexta-feira, 1º dia de maio de 1500" (I kiss the hands of your highness. From Porto Seguro, on your Island of the True Cross, today, Friday, 1 May 1500); see "Carta de Pero Vaz de Caminha," in Amado and Figueiredo, *Brasil 1500,* 117. Also see "Carta de D. Manuel aos Reis Católicos," in Amado and Figueiredo, *Brasil 1500,* 221; this letter is directed to the king and queen of Spain, Ferdinand and Isabella, and is dated 29 July 1501. The legend for Brazil on *Carta del Cantino* begins: "a vera cruz + chamada p. nome a quall achou pedralvres cabrall fidalgo da cassa del Rey de portugall"; see Cortesão and Mota, *Portugaliae Monumenta Cartographica,* 1:11.

10. There are striking similarities between *Carta del Cantino* and Caverio's *Planisphère nautique,* as noted in chapters 2 and 4.

11. Dürer, *Adam and Eve,* Cleveland Museum of Art, http://www.clevelandart.org/art /1944.473; P. J. Smith, *Zoology in Early Modern Culture,* 311.

12. *Arch of Honor,* ca. 1515–19, Metropolitan Museum of Art, New York, https://www .metmuseum.org/art/collection/search/388475?=&imgNo=16&tabName=related-objects. See also the "Original Woodblock for his woodcut (B. 159) with Arms of Michael Behaim (d. October 24, 1511) and autograph letter," B3 027 A04, Dept. of Drawings and Prints, Morgan Library and Museum, New York, https://www.themorgan.org/objects/item/214061; Panofsky, *Albrecht Dürer,* 1:45–47 and 62–64; Landau and Parshall, *Renaissance Print,* 42.

13. Trees of mastic (almaciga) and aloe (liñaloe), Lunes 5 Noviembre, in Colón, *Relaciones,* 56, and Columbus, *Diario,* 135. The pines: "Estando así dan voces los mozos grumetes, diciendo que vían pinales. Miró por la sierra, y vídolos tan guandes y tan

maravillosos, que no podía encarecer su altura y derechura como husos gordos y delgados, donde conosió que se podían hacer navíos é infinita tablazón y másteles para las mayores naos de España," Domingo 25 de Noviembre, in Colón, *Relaciones*, 74; English from Columbus, *Diario*, 171.

14. *OED Online*, s.v. "Brazil, n.1." Cantino (in Lisbon) to Ercole I, Duke of Ferrara, 17 October 1501, Chancelleria Ducale, Estero, Ambasciatori, Agenti e Corrispondenti Estensi, Fuori d'Italia, Spagna 1 (1468–1771) ASM 3988-74. For a Portuguese translation, see "Carta de Cantino a Hércules D'Este," in Amado and Figueiredo, *Brazil 1500*, 245–50. Vespucci (in Lisbon) to de' Medici, 1502: "Trovammovi infinito verzino, e molto buono, da caricarne quanti navilî oggi sono nel mare, e sanza costo nesuno," in Vespucci and Formisano, *Lettere di Viaggio*, 25; English translation from Vespucci and Formisano, *Letters from a New World*, 35.

15. On Newfoundland: Cantino (in Lisbon) to Ercole I, Duke of Ferrara, 17 October 1501, Chancelleria Ducale, Estero, Ambasciatori, Agenti e Corrispondenti Estensi, Fuori d'Italia, Spagna 1 (1468–1771) ASM 3988-74. This letter is transcribed in Harrisse, *Les Corte-Real*, 204–8. For a partial Portuguese translation of the letter, see "Carta de Cantino a Hércules D'Este," in Amado and Figueiredo, *Brazil 1500*, 245–50. Dispatch of Pietro Pasqualigo, 18 October 1501, to the Senate of Venice, as extracted from the *Diarii* of Sanuto, transcribed in Harrisse, *Les Corte-Real*, 212, and Pietro Pasqualigo (in Lisbon) to his Brothers, 19 October 1501, as transcribed in Harrisse, *Les Corte-Real*, 212. Legend on *Carta del Cantino*: "aquj ha muitos mastos," Cortesão and Mota, *Portugaliae Monumenta Cartographica*, 1:11.

16. Boehrer, *Parrot Culture*, 57; George, *Animals and Maps*, 60.

17. "Aqui he o Rey de menj o qual Rey he muoro e as gentes sam pretos e tratam muito com os navios que vam de portugall e com estas ilhas .s. ilha de sto thome etc. E daqui traem muytos escravos e ouro e algalia e outras cousas e papagayos pardos e buxios e pimeta"; Cortesão and Mota, *Portugaliae Monumenta Cartographica*, 1:12. *Papagayos pardos* can be translated as either "brown parrots" or, as I have chosen, "gray parrots." When used to describe skin color, *pardo* is often used to refer to a tone that is between white (*branco*) and black (*negro* or *preto*). The native peoples of Brazil are described by Caminha as "pardos maneira de avermelhados," meaning brown somewhat reddish. See "Carta de Pero Vaz de Caminha," in Amado and Figueiredo, *Brasil 1500*, 81.

18. "Carta de Pero Vaz de Caminha," in Amado and Figueiredo, *Brasil 1500*, 80–84. Cabral's parrot is described by Caminha as a "papagaio pardo."

19. Legend on *Carta de Cantino*: "Donde traçem ao muyto escelente principe dom manuell Rey de portugall cada anno doze caravelas com ouro traze cada caravera hua cõ outra xxv mjll pesos douro val cada peso qujnhentos rreaes e mais traem muytos escravos e pimento e outras cousas de muyto proueito"; Cortesão and Mota, *Portugaliae Monumenta Cartographica*, 1:12.

20. The Crown's share of the slave trade was at minimum an average of one thousand slaves per year while the private slave trade, whether by Portuguese merchants, Cape Verde islanders, or the São Tomean and Principé traders, was at minimum on average nearly two thousand slaves per year; see Elbl, "Volume of the Early Atlantic Slave Trade," 47, 59.

21. Devisse and Mollat, *From the Early Christian Era*, vol. 2, pt. 2 of Bindman and Gates, *Image of the Black*, 121–22, figs. 129, 130, 131, and 124; and 282, nn. 192, 196, and 197. Cresques (?), *Atlas de Cartes Marines*; Cattaneo, "Découvertes littéraires et géographiques," 95 and figure on p. 91-; Seville, *Etymologies*, 293 [XIV.v.ii–vi.i].

22. On Martin and Ludwig Schongauer and their engravings of elephants, see J. Hutchison, *Early German Artists: Martin Schongauer, Ludwig Schongauer, and Copyists*, in *The Illustrated Bartsch*, vol. 8, Commentary, pt. 1, pp. 253, 284–85. One of the twenty-five surviving copies of Martin Schongauer's *The Elephant* is at the Cleveland Museum of Art, https://www.clevelandart.org/art/1927.199; Ludwig Schongauer's engraving is reproduced in J. Hutchison, *Early German Artists*, 285, and survives in a single copy at the Albertina Museum, Vienna. On printed images on maps, and the nature of observation, see Dickinson, *German Masters*, 127–31. On Schongauer, Bartrum, *German Renaissance Prints*, 20.

23. Jane Hutchison, *Early German Artists*, 253, suggests that Martin Schongauer saw and carefully observed the elephant in Germany.

24. J. B. Harley, "New England Cartographies," in Harley and Laxton, *The New Nature of Maps*, 170–95 (quotation is from 188).

Chapter 6 · The Cannibal Scene

1. Harley, "Texts and Contexts," in Harley and Laxton, *New Nature of Maps*, 49. Davis identifies sixty-five cartographic works that depict cannibalism in northeastern South America; S. Davies, *Renaissance Ethnography*, 65. The literature on how cannibalism came to be associated with the Americas is vast; see, e.g., Arens, *The Man-Eating Myth*; Hulme, *Colonial Encounters*; Lestringant, *Cannibals*; Palencia-Roth, "Cannibalism and the New Man of Latin America"; and Whitehead, "Carib Cannibalism." Davies argues that the letters, accounts, and reports derived from Columbus's first voyages lie behind the first scenes of cannibalism on maps, and that they explain the construction of the imagery in *Weltkarte*; see *Renaissance Ethnography*, 65–88, esp. 84. In this chapter, I shall argue that although the term cannibalism can be linked to Columbus, the imagery is more closely associated with Vespucci.

2. *Portulan (Weltkarte)*, 1502–6, Cod.icon 133, Bayerische Staatsbibliothek, Munich. https://daten.digitale-sammlungen.de/~db/0000/bsb00003881/images/. The world chart measures 99 × 110.5 cm (about 3 ¼ × 3 ⅔ feet). On this world chart, see Kupčík, *Münchner Portolankarten / Munich Portolan Charts*, 28–34; he dates the chart to 1506. In "Amerigo Vespucci," I explore the possibility that this map was made by Vespucci. Bishop Carvajal, known for his learning, lived at the palace of the Mellini in Rome; see Weil-Garris and D'Amico, "The Renaissance Cardinal's Ideal Palace," 75, 101. His oration to Pope Alexander VI in June 1493, on behalf of Ferdinand and Isabella, mentioned the discovery of the Canary Islands as well as other unknown islands, presumably those discovered by Columbus, and it is one of the oldest documents on the discovery of the Americas; see Harrisse, *Description of Works*, 33–35. On his life, career, and influence, particularly in getting the papal bulls of 1493, see Goñi-Gaztambide, "Bernardino López de Carvajal."

3. Wood with Fels, *Power of Maps*, 122–24. Legend from *Weltkarte*: "Ista terra quae inventa fuit positum est nomen terra sanctae + eo quod in die sancte crucis inventa est et in ea est maxima copis ligni bresilli etiam invenitur cassia grossa ut brachium hominis, Aves Papagagi magni ut falcones et sunt rubri homines vero albi nullam legem habentes se invicem comedant."

4. *Woodcut of South American Indians. Di[s]e figur anzaigt vns das volck vnd in[s]el die gefunden i[s]t durch den c[r]i[s]tenlichen künig zü Po[r]tigal oder von [s]einen vnderthonen . . .*

printed in Augsburg in 1505 by Johann Froschauer, Spencer Collection, New York Public Library. The copy in the Bavarian State Library is *Dise figur anzaigt uns das volck und insel die gefunden ist durch den cristenlichen künig zu Portugal oder von seinen underthonen . . .* ca. 1503. Einbl. V,2, Bayerische Staatsbibliothek, Munich. https://bildsuche.digitale-sammlungen.de /index.html?c=viewer&bandnummer=bsb00058831&pimage=00001&lv=1&l=de. The woodcut has long been accepted as the first known image depicting cannibalism in the Americas; see Harrisse, *A Description of Works*, 51; Stevens, *Stevens's American Bibliographer*, 7–8; Schuller, "Oldest Known Illustration"; and S. Davies, *Renaissance Ethnography*, 79–81. On the provenance of the New York Public Library image, see Eames, *Description of a Wood Engraving*. The attribution of the broadside to Johannes Froschauer is made in several reference sources on early Americana, including Moraes, *Bibliographia Brasiliana*, 2:355, and Alden and Landis, *European Americana*, 1:9; see also Gier and Janota, *Augsburger Buchdruck und Verlagswesen*, 377.

5. Several well-known German masters worked in Augsburg in the first decade of the sixteenth century, including Hans Schäufelein, Jörg Breu, Daniel Hopfer, Leonard Beck, and Hans Burgkmair. The famous woodcutter Jost de Negker arrived in 1508, and several other woodcutters can be identified and named. On Augsburg as an early center of printing, see H. Lehmann-Haupt, "Book Illustration in Augsburg"; H. Graser and B. A. Tlusty, "Layers of Literacy," in Plummer and Charles, *Ideas and Cultural Margins*, 41; and Landau and Parshall, *Renaissance Print*, 7–15, 33–42, 174–79.

The German text below the image is: "Dise figur anzaigt uns das volck vnd insel die gefunden ist durch den christenlichen künig zü Portigal oder von seinen underthonen. Die leüt sind also nacket hübsch. braun wolgestalt von leib. ir heübter.halsz. arm. scham. füsz. frawen vnd mann ain wenig mit federn bedeckt. Auch haben die mann in iren angesichten vnd brust vid edel gestain. Es hat auch nyemantz nichts sunder sind alle ding gemain. Vnnd die mann habendt weyber welche in gefallen. es sey mütter, schwester oder freündt. darjnn haben sy kain vnderschayd. Sy streyten auch mit einander. Sy essen auch ainander selbs die erschlagen werden. vnd hencken das selbig fieisch in den rauch. Sy werden alt hundert vnd füntzig iar. Vnd haben kain regiment." English translation: "This figure represents to us the people and island, which have been discovered by the Christian King of Portugal or by his subjects. The people are thus naked, handsome, brown, well shaped in body, their heads, necks, arms, private parts, feet of men and women are a little covered with feathers. The men also have many precious stones in their faces and breasts. No one also has anything, but all things are in common. And the men have as wives those who please them, be they mothers, sisters, or friends, therein make they no distinction. They also fight with each other. They also eat each other even those who are slain, and hang the flesh of them in the smoke. They become a hundred and fifty years old. And have no government." From *Stevens's American Bibliographer*, 8.

6. On broadsides, see Park and Daston, "Unnatural Conceptions," 92; see also Daston and Park, *Wonders*, 173–214. Hall, "Before the Apocalypse," 12.

7. Strickland, "Sartorial Monsters of *Herzog Ernst*," 4; Dackerman, "Dürer's Indexical Fantasy," 164–71; and Daniel Zolli, "Cat.35. Albrecht Dürer," 172–75; quotation from p. in Dackerman, *Prints and the Pursuit of Knowledge*. The text above the woodcut states that the rhinoceros arrived in Lisbon in 1513, describes its color and hide, compares it to the elephant, and describes the enmity between it and the elephant—this last bit of information coming from the classical writer Pliny.

8. The Order of Christ (Ordem do Cristo) was created in Portugal in 1320 by transferring the properties of the Knights Templar to the new order. The Order of Christ became closely tied to the crown and was highy influential in overseas ventures; see Valente, "The New Frontier," and Olival, "The Military Orders."

9. Hulme, *Colonial Encounters* 15–16.

10. Zamora, *Reading Columbus*, 175. Classical accounts include Homer, *Odyssey*, IX; Herodotus, *History* IV.18, and Pliny, *Natural History* 6.20. Medieval tales include *Beowulf*, the *Chansons des gestes*, Chaucer, and *Herzog Ernst*; see Strickland, "The Sartorial Monsters of *Herzog Ernst*," 1. On medieval texts on monsters, see Daston and Park, *Wonders*, 25–26.

11. For monsters on maps, see Mittman, *Maps and Monsters*; C. Van Duzer, "Hic sunt dracones," in Mittman and Dendle, *The Ashgate Research Companion to Monsters*, 387–435; and Van Duzer "A Northern Refuge," 221–22. On travelers, see Daston and Park, *Wonders*, 30–39. Quotation from Marco Polo "Car il samblent des visages touz comme chiens maastins granz . . . et sont moult cruel gent, car il menjuent touz ceulz que il pueent prendre puis que il ne sont de leur gent. Et vivent de ris, de char et de lait," "Chapitre de l'ylle de Angamanam," in *Devisement du monde*, 6:22. Quotation from Palencia-Roth, "Mapping the Caribbean," 13. Lacarra, Ducay, and Frutos, *Libro del conosçimiento*.

12. Daston and Park, *Wonders*, 25–26; Mittman, *Maps and Monsters*, 39–58; Gog and Magog were believed to be wild peoples, who when released from their enclosed locations by the Antichrist, would attack and destroy Christendom. See Berenbaum and Skolnik, "Gog and Magog"; and Gow, "Gog and Magog." On the Hereford *Mappa Mundi*, Gog and Magog are confined to a walled peninsula, and an inscription describes them as anthropophagous peoples.

13. On Fra Mauro, see Scafi, "Mapping the End." For the *Nuremberg Chronicle*, see Schedel, *The Nuremberg Chronicle*, xii–xiii; Landau and Parshall, *Renaissance Print*, 38–42; Leitch, *Mapping Ethnography*, 26–30, and Scafi, *Mapping Paradise*, 166. On the border of the *mappamundi* in the *Nuremberg Chronicle*, the monsters are a six-armed man, a wild woman, a speechless man, a satyr, a hermaphrodite, a four-eyed Ethiopian, and a crane man; see Debra Higgs Strickland, "Map of the World," in "Foreign Bodies."

14. On Columbus on the lookout for monsters, see Zamora, *Reading Columbus*, 175. On the unreliability of Columbus, see Greenblatt, *Marvelous Possessions*, 7, 14, 55. According to Felipe Fernández-Armesto, the Santángel letter had eleven editions in 1493, most in Latin, while a German edition appeared in 1497; see introduction to Columbus, *Spanish Letter of Columbus*, viii. For a map showing the editions of the letter, as they were printed in Europe, see the Edney, *Geographical Diffusion of Columbus's First Letter*. A recent critical edition by Elizabeth Moore Willingham suggests that there were even more editions of the letter; see *The Mythical Indies*. On the importance of the letter for the dissemination of cannibalism in association with the Americas, see Hulme, *Colonial Encounters* 42–43. Quotations: "No hehallado onbres monstrudos commo muchos pensauan mas antes estoda gente demuy lindo acatamiento . . . ni son negros como em guinea saluo con sus cabelos corredios." Slightly later in the text: "Asique monstruos nohe hallado, ninoticia, saluo de vnaysla que es aqui enla segunda ala e[n]trada delas yndias que es poblada de vna iente que tie[n]en en todas las yslas por muy ferozes los quales comen carne vmana," as transcribed and translated by Elizabeth Willingham in *The Mythical Indies*. For a more modern transcription into Spanish, see Columbus, *The Spanish Letter of Columbus*, 19, 26.

15. Spanish for Las Casas's statement: "Este es el primer viaje y las derrotas y camino que hizo el almirante don xpoval lCristóbal] Colon quando d[e]scubrió las yndias puesto sumariamente sin el prólogo que hizo a los reyes, que va a la letra y comieça de esta ma[ne]ra." Columbus, *Diario*, 16. English translation from Columbus, *Diario*, 17. The complexity of this text, also known as *Relación del primer viaje de D. Cristóbal Colón*, is well known, because it is a summary of a copy, and Las Casas used it in his preparation of *Historia de las Indias*; see editor's introduction, Columbus, *Diario*, 3–14, and Zamora, *Reading Columbus*. As illustrated by Stephan Ruhstaller's study of the marginalia, Las Casas was especially interested in certain aspects of Columbus's account, critical of others, and aware of errors; see Ruhstaller, "Bartolomé de las Casas."

16. For these and subsequent quotations from the *Diario*, we shall use the English translations in Columbus, *Diario* and we shall use the Spanish from Colón, *Relaciónes*, which is based on Navarrete's transcription. Those wishing to see a facsimile of the original are directed to Columbus and Sanz, *Diario de Colón*. An exact transcription of the sixteenth-century text is available in Columbus, *Diario,* directly opposite the English translation. Spanish for first quotation: "y sobre este cabo encavalga otra tierra ó cabo que va también al Leste, á quien aquellos indios que llevaba llamaban *Bohio,* la cual decían que era muy grande y que había en ella gente que tenía un ojo en la frente, y otros que se llamaban Caníbales, á quien mostraban tener gran miedo. Y desque vieron que lleva este camino, diz que no podían hablar, porque los comían y que son gente muy armada. El Almirante dice que bien cree que había algo dello, mas que, pues eran armados, serían gente de razón, y creía que habrían captivado algunos y que porque no volvían á sus tierras dirían que los comían." Viernes, 23 de Noviembre, in Colón, *Relaciónes*, 72. English translation from Columbus, *Diario*, 167.

Spanish for second quotation: "Estimaba que la tierra que hoy vido de la parte Sudeste del *Cabo de Campana,* era la isla que llamaban los indios *Bohio*: parécelo por quel dicho cabo está apartado de aquella tierra. Toda la gente que hasta hoy ha hallado diz que tiene grandísimo temor de los Caniba ó Canima, y dicen que viven en esta isla de *Bohio,* la cual debe de ser muy grande, según le parece, y cree que van á tomar á aquellos á sus tierras y casas, como sean muy cobardes y no saber de armas. Y á esta causa le parecía que aquellos indios que traía no suelen poblarse á la costa de la mar, por ser vecinos á esta tierra, los cuales diz que después que le vieron tomar la vuelta de esta tierra no podían hablar temiendo que los habían de comer, y no les podía quitar el temor, y decían que no tenían sino un ojo y la cara de perro, y creía el Almirante que mentían, y sentía el Almirante que debían de ser del señorío del Gran Can, que los cautivaban." Lunes, 26 de Noviembre, in Colón, *Relaciónes*, 76–77. English translation from Columbus, *Diario,* 176–77.

17. Spanish for first quotation: "y decían que la *Isla de Bohio* era mayor que la *Juana* á que llaman *Cuba,* y que no está cercada de agua, y parece dar á entender ser tierra firme, que es aquí detrás de esta *Española,* á que ellos llaman *Caritaba,* y que es cosa infinita, y cuasi traen razón que ellos sean trabajados de gente astuta, porque todas estas islas viven con gran miedo de los de *Caniba,* y así torno á decir como otras veces dije, dice él, que *Caniba* no es otra cosa sino la gente del gran Can, que debe ser aquí muy vecino, y terná navíos y vernán a captivarlos, y como no vuelven creen que se los han comido." Martes, 11 de Diciembre, in Colón, *Relaciónes*, 96. English translation from Columbus, *Diario,* 216–17.

Spanish for second quotation: "Envió á pescar los marineros con redes: holgáronse mucho con los cristianos los indios y trujéronles ciertas flechas de los de Caniba ó de los

Caníbales, y son de las espigas de cañas, y exigiéronles unos palillos tostados y agudos y son muy largos. Mostráronles dos hombres que les faltaban algunos pedazos de carne de su cuerpo, y hiciéronles entender que los caníbales los habían comido á bocados: el Almirante no lo creyó." Lunes, 17 de diciembre, in Colón, *Relaciónes*, 105. English translation from Columbus, *Diario*, 237.

18. "Después que acabaron de comer llevó á la playa al Almirante, y el Almirante envió por un arco turquesco y un manojo de flechas, y el Almirante hizo tirar á un hombre de su compañía, que sabía dello; y el Señor, como no sepa qué sean armas, porque no las tienen ni las usan, le pareció gran cosa; aunque diz quel comienzo fué sobre habla de los de *Caniba*, quellos llaman *Caribes*, que los vienen á tomar, y traen arcos y flechas sin hierro, que todas aquellas tierras no había memoria dél, y de acero ni de otro metal, salvo de oro y de cobre, aunque cobre no había visto sino poco el Almirante. El Almirante le dijo por señas que los Reyes de Castilla mandarían destruir á los caribes, y que á todos se los mandarían traer las manos atadas." Miércoles, 26 de Diciembre, in Colón, *Relaciónes*, 128. English translation from Columbus, *Diario*, 285.

19. "Envió la barca á tierra en una hermosa playa para que tomasen de los ajes para comer, y hallaron ciertos hombres con arcos y flechas, con los cuales se pararon á hablar, y los compraron dos arcos y muchas flechas, y rogaron á uno de ellos que fuese á hablar al Almirante á la carabela; y vino, el cual diz que era muy disforme en el acatadura más que otros que hobiesen visto: tenía el rostro todo tiznado de carbón, puesto que en todas partes acostumbran de se teñir de diversos colores. Traía todos los cabellos muy largos y encogidos y atados atras, y despues puestos en una rebecilla de plumas de papagayos, y él así desnudo como los otros. Juzgó el Almirante que debía ser de los Caribes que comen los hombres." Domingo, 13 de Enero, in Colón, *Relaciónes*, 148. English translation from Columbus, *Diario*, 331.

20. "Dice más el Almirante, que en las islas pasadas estaban con gran temor de *Carib*, y en algunas le llamaban *Caniba*, pero en La Española *Carib*; y que debe de ser gente arriscada, pues andan por todas estas islas, y comen la gente que pueden haber." Domingo, 13 de Enero, in Colón, *Relaciónes*, 149. English translation from Columbus, *Diario*, 331.

21. "Volviéronse luego á la carabela los cristianos con su barca, y sabido por el Almirante, dijo que por una parte le había pesado y por otra no, porque hayan miedo á los cristianos, porque sin duda (dice él) la gente de allí es diz que de mal hacer, y que creía que eran los de *Carib*, y que comiesen los hombres," Domingo, 13 de Enero, in Colón, *Relaciónes*, 150–51. English translation from Columbus, *Diario*, 335.

22. First quotation: "Y que en la Isla de *Carib* había mucho alambre y en *Matinino*, puesto que será dificultoso en Carib, porque aquella gente diz que come carne humana." Martes, 15 de Enero, in Colón, *Relaciónes*, 152. English translation from Columbus, *Diario*, 339.

Second quotation: "Para ir, diz, que á la *Isla de Carib*, donde estaba la gente de quien todas aquellas islas y tierras tanto miedo tenían, porque diz que con sus canoas sin[ ]número andaban todas aquellas mares, y diz que comían los hombres que pueden haber." Miércoles, 16 de Enero, in Colón, *Relaciónes*, 154. English translation from Columbus, *Diario*, 341.

23. The Santángel letter was printed in Barcelona in 1493, the same year that the second voyage led by Columbus left Spain. It is possible Chanca had read the published letter, but he would have had far greater access to the oral knowledge circulating among

mariners. Spanish from Chanca letter: "Quatro o cinco huesos de braços é piernas de ombres. Luego que aquello vimos sospechamos que aquellas yslas eran las de Caribe, que son abitadas de gente que come carne umana, porque el Almirante por las señas que le avían dado del sitio destas yslas el otro camino los yndios de las yslas que antes avían descubierto, avía endereçado el camino por descubrirlas," in "Carta de Diego Alvarez Chanca," in Padron, *Primeras cartas*, 115.

24. The account of Niccolò Scillacio (also known as Nicolaus Syllacius) of Paiva is known as *De insulis meridiani atque Indici maris muper inventis;* it is dedicated to the Duke of Milan and dated 1494. In the dedication Scilliacio states that his information was drawn from letters written to him from Spain by Guillermo Como; see Syllacius, *De Insulis Meridiani*, 8. Original Latin: "Insulae canaballis parent: gens illa effera et indomita carnibus vescitur humanis: quos anthropophagos jure nuncupaverim"; and "devorant imbelles: a suis abstinent: parcunt cannaballis," Syllacius, *De Insulis Meridiani*, 26. In the Latin text, the term *canaballi* repeats, while Mulligan continues to use Carib in the English translation. In the Latin transcribed in Berchet, *Narrazioni Sincrone*, similarly the term is *cannibali*, especially 86–91. S. Davies, *Renaissance Ethnography*, 76, nn. 48, 49.

25. Silió, *La Carta de Juan de La Cosa*, 103.

26. Legends on *Carta Universal*: "Mar descubierto por los ingleses," and "Este cavo se descubrio en el año de Mil y IIII XCIX por Castilla syendo descubridor vicentinas," and "Ysla descubierta por Portugal"; see Pérez, "La Carta de Juan de la Cosa," 10–31. Legend on Caverio, *Planisphère nautique*: "Has Antilhas del Rey de Castella descoberta por Columbo [illeg.] almirante que es [illeg.] ditas insulas se descobriam per mandado do muito alto et poderoso principe Rey dom Fernando Rey de Castella" (The Antilles of the King of Castile discovered by Columbus . . . admiral who is . . . the said islands were discovered by order of the very great and powerful prince King Fernando of Castile). Legend on *Carta del Cantino*: "Has antillas del Rey de castella / descobertas por colombo almirante que es de las aqueles ditas ditas ilhas se descobriam por mandado do muyto alto e poderoso principe Rey dom Fernando Rey de Castella." Legend on *Universalis cosmographia*: "Iste insule per Columbum genuersem almirantem ex mandato regis Castelle invente sunt."

27. Legend for Brazil on *Carta del Cantino*: "A vera cruz + chamada p. nome a quall achou pedralvres caball fidalgo da cassa del Rey de portugall [que] elle a descobrio indo por capitamoor de quatorze naos que o dito Rey mandava a caliqut y enel caminho indo topou com esta terra em a qual terra se cree ser terra firme em a qual a muyta gente de descricam andam nuos omes y molheres como suas mais os pario sam mais brancos que bacos y tem os cabellos muyto corredios foy descoberta esta ditta terra em a era de quinhentos." As transcribed and translated in Cortesão and Mota, *Portugaliae Monumenta Cartographica*, 1:11.

28. "Capitaneo nauium quattuor decim: quas / rex Portugalie ad Calicutium misit / r ra [?] hic primum apparuit: que credebatur / firma cum reuera sit cum prius inuenta / parte circumflua mire sed non dum prorsus / cognite magnitudinis insula. In qua / virilis ac feminei etiam sexus homines / non aliter quam eos mater peperrit / ire asueuerunt. Et sunt hic quidem pau / lo albiores eis quos superiori nauiga / tione re mandato regis Castilie facta / reperiere." Special thanks to Tim O'Sullivan for helping me translate this legend.

29. Pérez, "La Carta de Juan de La Cosa," 26; Van Duzer, "A Northern Refuge," 222. Marco Polo described the inhabitants of Andaman Island as anthropophagous peoples, and

this would seem to be the source of the inscription on world charts and maps, such as *Carta del Cantino*, where the legend near Andaman states: "em estas ilha [*sic*] a gente della comense huns aos outros" (on these island [*sic*] the people eat each other) and further south "em estas tres [illeg.] ilhas nam ay nada senam gente muito pobre e nua" (On these three islands there is nothing except very poor and naked people). See also the transcriptions and translations in Cortesão and Mota *Portugaliae Monumenta Cartographica*, 1:12.

30. On Ruysch, see Shirley, *The Mapping of the World*, 25–27, and McGuirk, "Ruysch World Map: Census and Commentary." Johannes Ruysch, *Universalior cogniti orbis tabula ex recentibus confecta observationibus*, in Ptolemy, *Geographia* (Rome: Bernardinus Venetus de Vitalibus, 1508), James Ford Bell Library, University of Minnesota, Minneapolis, https://www.lib.umn.edu/bell/maps/ptolemy3.

31. "Passim incolitur haec regio: quae [a] plerisque alter terrarum orbis existimatur foeminae mares que vel nudi prorsus vel intextis radicibus aviumque pennis varii coloris ornati incedunt: vivitur multis in commune nulla religione nullo rege. Bella inter se continenter gerunt, humanaque capitivorum carne vescuntur. Aere adeo clementi utuntur ut supra annum 150 vivant. Raro aegrotant tuncque radicibus tantum curantur herbarum. Leones hic gignnuntur, serpentesque et aliae foedae belluae sylvae insunt montes fluminaque margaritarum atque auri maxima copie: avehuntur hinc a Lusitanis ligna brasi alias verzini et cassiae." Special thanks to Joan Burton for assistance with translating the Latin.

32. Moraes cites twelve or thirteen editions of the *Mundus Novus* in Latin, several in German and French, one in Dutch, and one in Czech, but none in Spanish or Portuguese. By 1550, it had appeared at least fifty times either on its own or in collections; see *Bibliographia Brasiliana*, 348. Latin text from Vespucci, *Mundus Novus* (1504): "Navigavimus autem secundum littus circa sexcentas leucas et sepe descendimus in terram et colloquebamur et conversabamur cum eaum regionum colonis et ab eis fraterne recipiebamur: et secum quandoque morabamur quindecim vel viginti dies continuos amicabiliter et hospitabilitur ut irferius intelliges." For a slightly different transcription, see Berchet, *Narrazioni Sincrone*, 126. English translation: "We sailed along the shore, approximately six hundred leagues, and we often landed and conversed with the inhabitants of those regions, and were warmly received by them, and sometimes stayed with them fifteen or twenty days at a time"; in Vespucci and Forminano, *Letters from a New World*, 48.

33. There were German editions of the *Mundus Novus* in both Nuremberg and Basel in 1505; see Moraes, *Bibliographia Brasiliana*, 355, but an earlier Latin edition at the John Carter Brown Library is from Augsburg 1504. On the connection between the broadside and the publication of the *Mundus Novus* letter, see Schuller, "The Oldest Illustration." Latin quotations in the text are from this edition, Vespucci, *Mundus Novus* (1504). First quotation: "Ubi vidi per domos humanam carnem salsam et contignationibus suspensam, uti apud nos moris est lardum suspendere et carnem suillam"; for a slightly different transcription, see Berchet, *Narrazioni Sincrone*, 128. English translation from Vespucci and Formisano, *Letters from a New World*, 50. Second Latin quotation: "Corpora enim habent magna, quadrata, bene disposita ac proportionate, et colore declinantia ad rubedinem . . . habent et comam amplam et nigram"; see also Berchet, *Narrazioni Sincrone*, 127. English translation from Vespucci and Formisano, *Letters from a New World*, 48. Third Latin quotation "Omnes utriusque sexus incedunt nudi: nullam corporis partem operientes"; see

also Berchet, *Narrazioni Sincrone*, 127; English translation from Vespucci and Formisano, *Letters from a New World*, 48.

34. On the *Lettera* as source for the image on the *Weltkarte*, see Anónimo (Kunstmann II)," in Sanz, *Bibliotheca Americana Vetustissima*, 1:261; W. Sturtevant, "First Visual Images," in Chiappelli, *First Images of America*, 1:420. Quote from Formisano, introduction to Vespucci and Formisano, *Letters from a New World*, xxiii–xxiv.

35. Rubiés, "Texts, Images and the Perception of 'Savages,'" 123–24; also S. Davies, *Renaissance Ethnography*, 81–83.

36. Vespucci claims to have kept a diary, but there is no such surviving document. A reference to a diary exists in Vespucci, *Mundus Novus* (1504): "Rerum notabilium Diarium feci"; see also Berchet, *Narrazioni Sincrone*, 134. English translation: "I kept a diary of the noteworthy things"; see Vespucci and Formisano, *Letters from a New World*, 55. On the familiar letters, see Formisano, introduction to Vespucci and Formisano, *Letters from a New World*, xxv–xxviii. Quotation from Gerbi, *Nature in the New World*, 36–37.

37. First quotation: "e questo è certo perchè trovammo nelle loro case la carne umana posta a·ffumo, e molta, e comprammo da loro x creature, sì maschi come femine, che stavano diliberati per il sagrificio"; Vespucci and Formisano, *Lettere di Viaggio*, 24; English translation from Vespucci and Formisano, *Letters from a New World*, 33. Second quotation: "E quando combattono, s'amazzano molto crudelmente, e quella parte che resta signor del campo, ttutti e' morti di loro bande li sotterano, e l'inimici li spezzano e se li mangiono; e quelli che pigliano l'imprigionano e·lli tengono per ischiavi alle loro case; e s'é femina, dormono con loro; e·sse è maschio lo maritano colle loro figliuole"; Vespucci and Formisano, *Lettere di Viaggio*, 24; English translation from Vespucci and Formisano, *Letters from a New World*, 33.

38. On the possibility that the voyage encountered the interpreters—the penal exiles left behind by Cabral—see Metcalf, *Go-betweens*, 38. Original text from Vespucci: "Quando li domandavamo che ci dicessino la causa, non sanno dare altra ragione, salvo che dicono *ab antico* cominciò infra loro questa maladizione, e vogliono vendicare la morte de' loro padri antipasati"; Vespucci and Formisano, *Lettere di Viaggio*, 24; English from Vespucci and Formisano, *Letters from a New World*, 34.

39. "E in certi tempi, quando vien loro una furia diabolica, convitano e' parenti, e 'l popolo, e·lle si mettono davanti—cioè la madre con tutti e' figliuoli che di lei n'ottiene—, e con certe cirimonie a saettate li amazzano e se li mangiano; e questo medesimo fanno a' detti schiavi e a' figliuoli che di lui nascono"; Vespucci and Formisano, *Lettere di Viaggio*, 24; English transation from Vespucci and Formisano, *Letters from a New World*, 33.

40. Subsequent accounts of anthropophagy by travelers, missionaries, and settlers in the sixteenth century, when read carefully, also describe ceremonial killings of enemies, usually prisoners of war, followed by rituals of anthropophagy. The act served both to ingest the strength of enemies and to avenge the deaths of one's own kin. On the existence of cannibalism in sixteenth-century Brazil, see Castro, *From the Enemy's Point of View*; Colin, "Woodcutters and Cannibals"; Duffy and Metcalf, *Return of Hans Staden*; Forsyth, "Three Cheers"; Gareis, "Cannibals, *Bons Sauvages*"; Klarer, "Cannibalism and Carnivalesque"; Lestringant, *Cannibals,* 51–80; Fernandes, *Organização social dos Tupinambá*; Martel, "Hans Staden's Captive Soul"; Villas-Bôas, "Anatomy of Cannibalism"; Staden, Whitehead, and Harbsmeier,

Hans Staden's True History, xxi–civ; Whitehead, "Hans Staden and the Cultural Politics"; and Zika, "Cannibalism and Witchcraft."

41. Original text in Soderini letter: "Stavano già le donne faccendo pezzi del cristiano . . . mostrandoci molti pezzi e magiandoseli"; Vespucci and Formisano, *Lettere di Viaggio,* 60; English translation from Vespucci and Formisano, *Letters from a New World,* 89. Suzi Colin argues that the vividness of the image of cannibalism on *Weltkarte* suggests that it was drawn by an eyewitness but concludes that the illuminator "based his portrayal on accounts that had been embellished by various intermediaries"; "Woodcutters and Cannibals," 176. Surekha Davies dismisses even more strongly any relationship between the account in the *Lettera* and the image on *Weltkarte*; *Renaissance Ethnography,* 86, n. 83.

42. On the use of cannibalism to justify enslavement, see Whitehead, "Carib Cannibalism"; E. Stone, "Chasing 'Caribs,'" in Fynn-Paul and Pargas, *Slaving Zones,* 118–47; and Metcalf, *Go-betweens,* 173–80.

43. See S. Davies, *Renaissance Ethnography,* 101–4, and Colin, "Woodcutters and Cannibals," 176. Martin Waldseemüller, *Carta Marina Navigatoria Portugallen Navigationes Atque Tocius Cogniti Orsis Terre Marisque* ([Strasbourg?], 1516). Library of Congress, Washington, DC. Lorenz Fries, *Carta Marina.* 1530. Museum zu Allerheiligen Schaffhausen, Switzerland. In *Cannibals,* Lestringant traces the degrading of the "other" in the changing portrayal of cannibals from the writings from the Renaissance to the Romantic period.

Conclusion

1. Saraiva quotation from Domingues, "Colombo," 106; Cortesão and Mota, "Introduction to Volume I," *Portugaliae Monumenta Cartographica,* 1:xlv; T. Campbell, "Portolan Charts," 372–73.

2. Wood with Fels, *The Power of Maps,* 4; Wood makes the same point in very similar language in Wood with Fels and Frygier, *Rethinking the Power of Maps,* 15. Skelton also noted the tragic disappearance of maps; see Karrow, "Centers of Map Publishing," 612.

3. Domingues, "Colombo," 106; Cortesão, *A política de sigilo*; and "Siglio, Política de," in Albuquerque and Domingues, *Dicionário de Historia,* 2:989–92. In English, see Kimble, "Portuguese Policy," 653–59. Following Foucault's arguments about power and knowledge, Harley saw the growing secrecy surrounding maps in the sixteenth century as indicative of the determination of the state, with its monopolistic commercial interests, to keep cartographic knowledge secret; see "Silences and Secrecy," 60–63. On critiques of the theory of secrecy, see Domingues, "Colombo," 110–11; Albuquerque, *Historia,* 255; and Alegria, Garcia, and Relaño, "Cartografia e viagens," 38–39. For other explanations for the loss of charts, see Marques, *A cartografia dos descobrimentos,* 38; and Astengo, "Renaissance Chart Tradition," 177–78, who suggests that most charts were destroyed by wear and tear, though later events, such as the bombing raids of World War II, also took their toll.

4. Campbell writes that only a minute fraction of portolan charts produced in the fourteenth and fifteenth centuries survives; see "Portolan Charts," 372–73. This point is also made for the Portuguese charts by Cortesão, *Cartografia portuguesa antiga,* 171–73. Quotations from Marques, *A cartografia dos descobrimentos,* 38; and Alegria et al., "Portuguese Cartography in the Renaissance," 983.

5. Jacob, *The Sovereign Map*, 33–46; quotations from pp. 35, 34, and 33.

6. Jacob's definition of monumental concrete maps does imply ephemerality, for he recognizes that monumental maps are "often drawn on a fragile material, vulnerable to an accidental disappearance," *Sovereign Map*, 33. Jacob also sees manuscript maps as ephemeral: "The era of manuscript maps was one of prototypes and, in a certain fashion, of ephemeral maps," and he sees the printing press as assuring "the safeguarding and the permanence of maps belonging to the past" (63). However, as I argue, printed maps better fit the modern definition of ephemera, whereas charts are better seen as examples of proto-ephemera.

7. *OED Online* s. vv. "ephemeron, n."; "ephemera, n.2."; J. Lewis, *Printed Ephemera*; Rickards and Twyman, *Encyclopedia of Ephemera*, v; Dann, "Ephemera Collecting"; Barlow, "Advertising Ephemera."

8. C. Smith, "Map Ownership," 77; Carlton, *Worldly Consumers*, 58–60; 101–2; on maps as ephemera, see McKinstry, "What Is Ephemera—Maps," and my, "Maps as Ephemera" and "Ephemera or Ephemeral?" The failure of ephemera to make their way into archives was the theme of the conference, "Ephemera and Archive," organized by Tani Barlow and Steve Lewis, Rice University, 2–4 December 2011. T. Schlereth, "Twentieth-Century Highway Maps," in Buisseret, *From Sea Charts to Satellite Images*, 264.

9. Astengo, "Renaissance Chart Tradition," 236–37.

10. On the loss of printed maps, see Grenacher, "The Woodcut Map," 31. Quotations by Campbell from T. Campbell and Destombes, *The Earliest Printed Maps*, 16. Koeman, *History of Abraham Ortelius*, 24; Jacob, *The Sovereign Map*, 33–34.

11. Colin, "Woodcutters and Cannibals," 178; Johnson, *Carta Marina*, 21.

12. Carlton, *Worldly Consumers*; C. Smith, "Map Ownership."

13. Carlton, *Worldly Consumers*, 59–60.

14. John W. Hessler, "Carta Marina," in Dunkelman, *The Jay I. Kislak Collection*, 105.

15. In one of the text boxes of the Carta Marina of 1516, Waldseemüller states that one thousand copies of a previous map had been printed. Presumably, although not definitively, this refers to the 1507 wall map; see Hessler "Translation of Large Text."

16. On the *Sammelband*, see Hessler, *Renaissance Globemaker's Toolbox*; bookplate quotation, 35. In other writings, Schöner was concerned that scientific knowledge could be lost or repressed; see Hessler, *Renaissance Globemaker's Toolbox*, 35–40.

17. Hessler, *Renaissance Globemaker's Toolbox*; Harrisse, *Discovery of North America*, 278; 443–45; Waldseemüller, Fischer, and Wieser, *Älteste Karte*.

18. On the acquisition of Waldseemüller's 1507 maps, see "Library of Congress Acquires Only Known Copy"; Hayes, "A Million-Dollar Map"; and Hebert, "The Map that Named America." On the acquisition of the *Carta Marina* by Jay Kislak, see Hagerty, "Real-Estate Mogul." On the discovery of a presumed fifth set of globe gores, see Labi, "An American Treasure Turns Up." A presumed sixth set turned up in 2017 and was to be auctioned on 13 December 2017; see Crowe, "More on the Waldseemüller Globe." On the globe gores being forgeries, see Jackson, "How a Rare Map."

19. Astengo argues that the flow of information dried up for Mediterranean chartmakers in the sixteenth and seventeenth centuries and that their charts and atlases used information that was increasingly out of date. Manuscript charts were eventually replaced by printed charts and printed atlases; see "Renaissance Chart Tradition," 236–37.

20. Quinn, "Artists and Illustrators," 245–46. See the perceptive essays collected by David Woodwart in *Art and Cartography*, as well as Carlton, *Wordly Consumers*, and Buisseret, *Mapmakers' Quest*.

21. Milano and Battini, *Mapamundi Catalán Estense*, 14; L. Chiappini, "1598: Chronicle of a Changement."

22. Tuohy, *Herculean Ferrara*, 2–4; Cortesão and Mota, "Cantino Planisphere," in Cortesão and Mota, *Portugaliae Monumenta Cartographica*, 1:7; Ricci, *Carte geografiche miniate*.

23. On the damage during World War II, see Milano and Battini, *Mapamundi Catalán Estense,* 18; "Works of Art in Italy"; and the application for World Heritage listing, "World Heritage List, Ferrara," 39. On the earthquake in 2012, see "Avviso Agli Utenti." On the recent digitization, see Estense Digital Library, "Digitalizzazione Carta del Cantino," 13 December 2019, https://www.progetto.estense-digital-library.it/digitalizzazione-della -carta-del-cantino/.

24. Barlow, "Advertising Ephemera." For Barlow, these meanings are linked to modernity, consumption, and coded sexuality.

25. Carlton, *Wordly Consumers*, 59–62.

26. Buisseret, *Monarchs, Ministers, and Maps*, 1; Gerard Mercator, *Nova et aucta orbis terrae descriptio ad usum navigantium emendata accommodata*, 1569, Département Cartes et plans, GE A-1064 (RES), Bibliothèque nationale de France, http://gallica.bnf.fr/ark:/12148 /btv1b7200344k. Legend on cannibalism: "Indigenae passim per Indiam novam sunt antropophagi" (The natives of various parts of the New Indies are cannibals).

27. On Mercator, see Brotton, *A History of the World,* 218–59, and Taylor, *The World of Gerard Mercator*. Mercator's projection allowed the lines of constant compass bearing (also known as rhumbs or loxodromic lines) to appear straight on a map or chart. This allowed navigators to plot their course directly onto a chart. On the controversy over the Mercator projection, see Wood with Fels, *Power of Maps*, 57–61.

28. Wood with Fels and Krygier, *Rethinking the Power of Maps*, 27; 27–343. Wood writes that nation-states were *"very hard to imagine without the creative intercession of the map,"* 34 (emphasis in the original). Harley, "Silences and Secrecy," 59, and Harley, "Maps, Knowledge, and Power," 79, 60, 57–58, both in Harley and Laxton, *The New Nature of Maps*.

29. Harley, "Texts and Contexts," 40, and Harley, "Maps, Knowledge, and Power," 63, both in Harley and Laxton, *The New Nature of Maps*, 63.

Maps and Images

Aguiar, Jorge de. *Portolan Chart of the Mediterranean Sea, the North Atlantic Ocean, the Black Sea, and the West African coast as far south as Sierra Leone.* 1492. 30cea/1492. Beinecke Rare Book and Manuscript Library, Yale University, New Haven, CT. http://brbl-dl.library.yale.edu/vufind/Record/3433718.

Alte Welt. Ca. 1500. Cod.icon. 140. Bayerische Staatsbibliothek, Munich.

Arch of Honor. Ca. 1515–19. Metropolitan Museum of Art, New York. https://www.metmuseum.org/art/collection/search/388475?=&imgNo=16&tabName=related-objects.

Barbari, Jacopo de'. *View of Venice.* 1500. Cleveland Museum of Art. http://www.clevelandart.org/art/1949.565.1.

———. *Woodcut Blocks for Veduta di Venezia.* 1500. Museo Correr, Venezia.

Becharius, Franciscus. *Portolan chart of the Mediterranean Sea, the North Atlantic Ocean, the Black Sea, and the northwestern African coast.* [1403.] Art Storage 1980 158. Beinecke Rare Book and Manuscript Library, Yale University, New Haven, CT. https://brbl-dl.library.yale.edu/vufind/Record/3521236.

Behaim-Globus. Germanisches National Museum, Nuremberg. https://www.gnm.de/index.php?id=749&L=1.

Benincasa, Grazioso. *A Portolan Chart of Europe.* 1470. Additional MS, 31318A, British Library, London. https://www.bl.uk/collection-items/a-portolan-chart-of-europe-by-grazioso-benincasa-with-the-aegean.

Bianco, Andrea. *L'Atlante.* Ms. It. Z, 76 (= 4783), Biblioteca Nazionale Marciana, Venezia. https://www.movio.beniculturali.it/bnm/ridottiprocuratorisanmarco/it/118/andrea-bianco.

Buondelmonti, Cristoforo, Henricus Martellus Germanus, and Evelyn Edson. *Description of the Aegean and Other Islands.* Facsimile Edition. New York: Italica Press, 2018.

Carta del Cantino. 1502. Gallerie Estensi, Biblioteca Estense Universitaria, Modena. https://www.gallerie-estensi.beniculturali.it/opere/carta-del-cantino/.

Carta nautica: Francia, Spagna, Africa occidentali. 1471–82. Biblioteca Estense Universitaria, Modena. https://www.gallerie-estensi.beniculturali.it/opere/carta-nautica-francia-spagna-africa-occidentali/.

Caverio, Nicolay de. *Planisphère nautique / Opus Nicolay de Caverio ianuensis*. 1506. Département Cartes et plans, GE SH ARCH-1. Bibliothèque nationale de France, Paris. http://gallica.bnf.fr/ark:/12148/btv1b550070757.

Cosa, Juan de la. *Carta Universal de Juan de la Cosa*. MNM-257. Museo Naval de Madrid & Biblioteca Virtual del Ministerio de Defensa, Madrid. http://bibliotecavirtualdefensa.es/BVMDefensa/i18n/consulta/registro.cmd?id=16822.

Cresques, Abraham (?). *Atlas de cartes marines, dit [Atlas catalan]*. Manuscript. Département des Manuscrits. Espagnol 30. Bibliothèque nationale de France. https://gallica.bnf.fr /ark:/12148/btv1b55002481n.

Dise figur anzaigt uns das volck und insel die gefunden ist durch den cristenlichen künig zu Portigal oder von seinen underthonen . . . ca. 1503. Bayerische Staatsbibliothek, Munich. https://bildsuche.digitale-sammlungen.de/index.html?c=viewer&bandnummer =bsb00058831&pimage=00001&lv=1&l=de.

Dürer, Albrecht. *Adam and Eve*, 1504. Cleveland Museum of Art. http://www.clevelandart .org/art/1944.473.

———. *Original Woodblock for His Woodcut (B. 159) with Arms of Michael Behaim (d. October 24, 1511) and Autograph Letter*. Germany, 1500. J. P. Morgan Library, New York.

Edney, Matthew H. *Geographical Diffusion of Columbus's First Letter*. 1996. https:// oshermaps.org/special-map-exhibits/columbus-letter/iv-diffusion-columbuss-letter -through-europe-1493-1497.

Fries, Lorenz. *Carta Marina*. 1530. Museum zu Allerheiligen Schaffhausen. Switzerland.

Froschauer, Johann. *Woodcut of South American Indians. Di[s]e figur anzaigt vns das volck vnd in[s]el die gefunden i[s]t durch den c[r]i[s]tenlichen künig zü Po[r]tigal oder von [s]einen vnderthonen . . .* Augsburg, 1505. Spencer Collection, New York Public Library.

———. *Dise figur anzaigt uns das volck und insel die gefunden ist durch den cristenlichen künig zu Portigal oder von seinen underthonen . . .* ca. 1503. Einbl. V,2. Bayerische Staatsbibliothek, Munich. https://bildsuche.digitale-sammlungen.de/index.html?c =viewer&bandnummer=bsb00058831&pimage=00001&lv=1&l=de.

Germano, Nicolò. *Planisfero*. 1466. In Ptolemy, *Cosmographia*. 1466. Lat. 463 = alfa.X.1.3, Biblioteca Estense Universitaria, Modena. https://www.gallerie-estensi.beniculturali.it /opere/cosmografia-di-tolomeo/.

Hammer, Heinrich [Henricus Martellus]. *Map of the world of Christopher Columbus*. Ca. 1489. Beinecke Rare Book and Manuscript Library, Yale University, New Haven, CT. https://brbl-dl.library.yale.edu/vufind/Record/3435243.

Holbein, Hans. *"The Ambassadors" Globe*. 1885? https://brbl-dl.library.yale.edu/vufind /Record/3486397.

Homem, Lopo. *Atlas nautique du Monde, dit [atlas Miller]*. 1519. Département Cartes et plans, GE DD-683 (5 RES), Bibliothèque nationale de France, Paris. https://gallica.bnf.fr /ark:/12148/btv1b55002607s.

Homem, Lopo, Jorge Reinel, Pedro Reinel, and António de Holanda. *Atlas Miller* [1519]. Facsimile edition. Barcelona: M. Moleiro Editor, S.A, 2003.

Mappamondo Catalano estense. 1450–60. C.G.A. 1. Biblioteca Estense Universitaria, Modena. https://www.gallerie-estensi.beniculturali.it/opere/carta-catalana/.

Mappa Mundi. Hereford Cathedral. https://www.themappamundi.co.uk/index.php.

Martellus, Henricus. *Insularium Illustratum*. Pelagios Project: Digitised Insularium Illustratum, 2017. Additional MS 15760, British Library, London. http://data.bl.uk/pelagios/pelo2 .html.

Mauro, Fra. *Il Mappamondo di Fra Mauro*. 1450. Biblioteca Nazionale Marciana, Venice. https://marciana.venezia.sbn.it/la-biblioteca/il-patrimonio/patrimonio-librario/il -mappamondo-di-fra-mauro.

Mercator, Gerard. *Nova et aucta orbis terrae descriptio ad usum navigantium emendata accommodata*, 1569. Département Cartes et plans, GE A-1064 (RES), Bibliothèque nationale de France, Paris. http://gallica.bnf.fr/ark:/12148/btv1b7200344k.

Portulan (Weltkarte). 1502–6. Cod.icon 133. Bayerische Staatsbibliothek, Munich. https:// daten.digitale-sammlungen.de/~db/0000/bsb00003881/images/.

Portolan Chart (King-Hamy). HM 00045. The Huntington Library, San Marino, CA. http://dpg.lib.berkeley.edu/webdb/dsheh/heh_brf?Description=&CallNumber=HM+45 and http://ds.lib.berkeley.edu/HM00045_43.

Portolan chart of the Mediterranean Sea. Ca. 1320–50. Geography and Map Division, G5672.M4P5 13—.P6, Library of Congress, Washington, DC. http://hdl.loc.gov/loc .gmd/g5672m.ct000821.

Ptolemy, Claudius. *Claudii Ptolemei Viri Alexandrini Mathematice Discipline Philosophi Doctissimi Geographie Opus Nouissima Traductione e Grecorum Archetypis Castigatissime Pressum, Ceteris Ante Lucubratorum Multo Prestantius . . .* Strasbourg, 1513. Library of Congress, Washington, DC. https://www.loc.gov/item/48041351/.

———. *Cosmographia*. 1466. Lat. 463=alfa.X.1.3. Biblioteca Estense Universitaria, Modena. http://bibliotecaestense.beniculturali.it/info/img/geo/i-mo-beu-alfa.x.1.3.html.

———. *Cosmographia*. Ulm: Johann Reger, 1486. Stanford University Library, Stanford, CA. https://purl.stanford.edu/fs844yc9264.

———. *L'atlante di Borso d'Este: La Cosmographia di Claudio Tolomeo della Biblioteca Estense Universitaria di Modena*. Modena: Il Bulino, 2006.

Ptolemy, Claudius, and R. A. Skelton. *Geographia: Strassburg, 1513*. Amsterdam: Theatrum Orbis Terrarum, 1966.

Ptolemy, Claudius, and Edward Luther Stevenson. *Geography of Claudius Ptolemy*. Florence, 1930.

Reinel, Pedro. *Portulan (Atlantik)*. Ca. 1504. Cod.icon. 132. Bayerische Staatsbibliothek, Munich.https://daten.digitale-sammlungen.de/~db/0000/bsb00002580/images/.

———. *Portulan. Carte des côtes de l'Europe et de l'Afrique occidentales, dressée par Pedro Reinel vers 1485*. 2 Fi 1582 bis. Archives départementales de la Gironde, Bordeaux.

Rosselli, Francesco. *World Map*. 1508. G201:1/53. National Maritime Museum, Greenwich. https://collections.rmg.co.uk/collections/objects/244434.html.

Rosselli, Francesco di Lorenzo, and Giovanni Matteo Contarini. *Mundu [sic] spericum . . . cognosces diligentia joani mathei Contareni, arte et ingenio francisci Roselli florentini 1506 notu*. 1506. Cartographic Items Maps C.2.cc.4. British Library, London. https://www.bl .uk/collection-items/first-known-printed-world-map-showing-america.

Rotz, Jean, and Helen Wallis. *The Maps and Text of the Boke of Idrography Presented by Jean Rotz to Henry VIII: Now in the British Library*. Facsimile edition. Oxford: Oxford University Press, 1981.

Rüst, Hans. *Das ist die mapa mu[n]di un[d] alle land un[d] kungk reich wie sie ligend in der ga[n]sze welt.* Augsburg, ca. 1480. Incunable Collection, PML 19921. Morgan Library and Museum, New York. https://www.themorgan.org/incunables/145336.

Ruysch, Johannes. *Universalior cogniti orbis tabula ex recentibus confecta observationibus.* In Ptolemy, *Geographia.* Rome: Bernardinus Venetus de Vitalibus, 1508. James Ford Bell Library, University of Minnesota, Minneapolis. https://www.lib.umn.edu/bell/maps /ptolemy3.

————. *Universalior cogniti orbis tabula ex recentibus confecta observationibus.* In Ptolemy, *Geographia.* Rome: Bernardinus Venetus de Vitalibus, 1508. John Carter Brown Library, Brown University, Providence, RI. https://jcb.lunaimaging.com/luna/servlet/detail/JCBM APS~1-1-1716-103410003:Universalior-cogniti-orbis-tabula-e?qvq=q:Ruysch&mi=0&trs=1.

Schnitzer, Johannes. *Mappamundi.* In Claudius Ptolemy, *Cosmographia.* Ulm: Johann Reger, 1486. James Ford Bell Library, University of Minnesota, Minneapolis. http://gallery.lib .umn.edu/archive/original/762c5731e77392a41396f74adb9dcc20.jpg.

Schöner, Johann. *Fragmentary terrestrial globe gores.* St. Dié, [1517]. Library of Congress, Washington, DC.

————. *Globe Recreated from Facsimiles of Terrestrial Globe Fragments from the Schöner Sammelband.* N.d. Library of Congress, Washington, DC. https://www.loc.gov/item/2016586441/.

Schongauer, Martin. *The Elephant.* Ca. 1435–91. Metropolitan Museum of Art, New York. https://www.metmuseum.org/art/collection/search/336141.

Vallseca, Gabriel. *Carte marine de la mer Méditerranée et de la mer Noire*, Gabriel Devallsecha la affeta en mallorcha an MCCCCXXXXVII. Département Cartes et plans, GE C-4607 (RES), Bibliothèque nationale de France, Paris. https://gallica.bnf.fr/ark:/12148 /btv1b53064893j.

Viladestes, Mecia de. *Carte marine de l'océan Atlantique Nord-Est, de la mer Méditerranée, de la mer Noire, de la mer Rouge, d'une partie de la mer Caspienne, du golfe Persique et de la mer Baltique*, 1413. Cartes et plans, GE AA-566 (RES), Bibliothèque nationale de France, Paris. https://gallica.bnf.fr/ark:/12148/btv1b55007074s.

Waldseemüller, Martin. *Carta Marina Navigatoria Portugallen Navigationes Atque Tocius Cogniti Orsis Terre Marisque.* [Strasbourg?], 1516. Library of Congress, Washington, DC.

————. *Globe Gores.* Saint-Dié, France: Walter and Nikolaus Lud?, 1507. James Ford Bell Library, University of Minnesota, Minneapolis.

————. *Universalis Cosmographia Secundum Ptholomaei Traditionem et Americi Vespucii Alioru[m]que Lustrationes.* 1507. Library of Congress, Washington, DC. https://www.loc .gov/resource/g3200.ct000725C/.

Historical Texts

Academia Real das Sciencias. *Collecção de noticias para a historia e geografia das nações ultramarinas que vivem nos dominios portuguezes, ou lhes são visinhas.* 3 vols. Lisbon: Academia Real das Sciencias, 1812.

Alberti, Leon Battista, and Lodovico Domenichi. *La pittura.* In *Vinegia: Appresso Gabriel Giolito de Ferrari*, 1547. http://archive.org/details/gri_pitturexxxxx00albe.

Amado, Janaína, and Figueiredo, Luiz Carlos. *Brasil 1500: Quarenta documentos.* São Paulo and Brasília, DF: Imprensa Oficial SP & Editora UnB, 2001.

Anghiera, Pietro Martire d'. *De orbe novo decades*. Alcalá: Arnaldi Guillelmi, 1516.

——. *De Orbe Novo de Pierre Martyr Anghiera: Les huit décades*. Translated by Paul Gaffarel. Paris. E. Leroux, 1907.

——. *De Orbe Novo, the Eight Decades of Peter Martyr d'Anghera*. Translated by Francis Augustus MacNutt. New York: G. P. Putnam's Sons, 1912.

Anonymous. "Relação do Piloto Anônimo." In *Brasil 1500: Quarenta documentos*, edited by Amado, Janaína and Luiz Carlos Figueiredo, 131–40. São Paulo, SP, and Brasília, DF: Imprensa Oficial SP & Editora UnB, 2001.

Barros, João de. *Asia de João de Barros: dos feitos que os portugueses fizeram no descobrimento e conquista dos mares e terras do oriente. Primeira Década*. Edited by António Baião. 4th ed. Coimbra: 1932; rpt., Lisbon: Impr. Nacional-Casa da Moeda, 1988.

——. *Asia de Joam de Barros dos fectos que os Portugueses fizeram no descobrimento & conquista dos mares & terras do Oriente. Primeira Decada*. Lisbon: 1553.

Berchet, Gugliemo. *Fonti Italiane per la Storia della Scoperta del Nuovo Mondo*. In *Raccolta di Documenti e Studi. Pubblicati dalla R. Commissione Colombiana pel quarto centenario dalla scoperta dell'America*. Part III: I. Rome: Auspici il Ministero della Pubblica Instruzione, 1892.

——. *Narrazione Sincrone*. In *Raccolta di Documenti e Studi. Pubblicati dalla R. Commissione Colombiana pel quarto centenario dalla scoperta dell'America*. Part III: II. Rome: Auspici il Ministero della Pubblica Instruzione, 1893.

Cantino, Alberto. "Letters." Chancelleria Ducale, Estero, Ambasciatori, Agenti e Corrispondenti Estensi, Fuori d'Italia—Spagna. Archivio di Stato di Modena.

Castillo, Bernal Diaz del. *Historia verdadera de la conquista de la Nueva España*. Mexico City: Editorial Porrúa, 1966.

Chaves, Alonso de. "Quatri partitu en cosmographia pratica i por otro no[m]bre llamado Espeio de Navegantes," Ca. 1537. Real Academia de la Historia and Biblioteca Digital Real Adademia del la Historia, Madrid.

Colón, Cristóbal. *Relaciones y cartas*. Madrid: La Viuda de Hernando, 1892.

Columbus, Christopher. *The Diario of Christopher Columbus' First Voyage to America, 1492–1493. Abstracted by Fray Bartolomé de las Casas*. Transcribed and translated by Oliver Dunn and James E. Kelly Jr. Norman: University of Oklahoma Press, 1989.

——. *The Spanish Letter of Columbus: A Facsimile of the Original Edition Published by Vernard Quaritch in 1891*. Edited by Anthony Payne. London: Quaritch, 2006.

Cortés, Martín. *Breve compendio de la sphera y de la arte de navegar: con nuevos instrumentos y reglas, exemplificado con muy subtiles demonstraciones*. Seville: Anton Alvarez, 1551.

Cortés, Martín, and Eden, Richard. *The Arte of Nauigation Conteyning a Compendious Description of the Sphere . . .* London: Abell Jeffes and Richard Watkins, 1589.

Dürer, Albrecht, and William Martin Conway. *The Writings of Albrecht Dürer*. New York: Philosophical Library, 1958.

Dürer, Albrecht, and Erwin Panofsky. *Dürer's Apocalypse*. London, Eugrammia Press, 1964.

Faleiro, Francisco. *Tratado del esfera y del arte de marear*. Seville, 1535.

Figueiredo, Manoel de. *Hydrographia, exame de pilotos, no qual se contem as regras que todo piloto deve guardar em suas navegações. . . .* Lisbon: Vicente Alvarez, 1614.

Greenlee, William Brooks. *The Voyage of Pedro Alvares Cabral to Brazil and India, from Contemporary Documents and Narratives*. London: Hakluyt Society, 1938.

Herodotus. *The History.* Translated by David Green. Chicago: University of Chicago Press, 1987.

Herrera, Antonio. *Historia general de los hechos de los Castellanos en las Islas i Tierra Firme del Mar Oceano. Decada Terzera.* Madrid: En la emplenta Real, 1601.

Lacarra, María Jesús, María Carmen Lacarra Ducay, and Alberto Montaner Frutos. *Libro del conosçimiento de todos los regnos et tierras et señorios que son por el mundo, et de las señales et armas que han.* Facsimile edition. Zaragoza: Institución "Fernando el Católico," 1999.

Lamb, Ursula. *The Quarti Partitu en Cosmographia by Alonso de Chaves: An Interpretation.* Coimbra: Junta de Investigações do Ultramar, 1969.

Las Casas, Bartolomé de. *Historia de las Indias.* Vols. 1–3. Madrid: Impr. de M. Ginesta, 1875.

Las Casas, Bartolomé de, and Geoffrey Symcox. *Las Casas on Columbus: The Third Voyage.* Turnhout: Brepols, 2001.

Montalboddo, Fracanzano da, Amerigo Vespucci, and Arcangelo Madrignani. *Itinerariu[m] Portugalle[n]siu[m] e Lusitania in India[m] [et] inde in occidentem [et] demum ad aquilonem.* Milan, 1508.

Navarrete, Martín Fernández de. *Colección de los viages y descubrimientos que hicieron por mar los españoles desde fines del siglo XV. Con varios documentos inéditos concernientes á la historia de la marina castellana y de los establecimientos españoles en Indias.* Vol. 2: *Documentos de Colon y de las primeiras poblaciones.* Madrid: Imprenta real, 1825.

———. *Colección de los viages y descubrimientos que hicieron por mar los españoles desde fines del siglo XV: con varios documentos inéditos concernientes á la historia de la marina castellana y de los establecimientos españoles en Indias.* Vol. 3: *Viages menores, y los de Vespucio; poblaciones en el Darien, suplemento al tomo II.* Madrid: Imprenta real, 1829.

———. *Coleccion de los viages y descubrimientos que hicieron por mar los espanoles: desde fines del siglo XV, con varios documentos ineditos conciernentes a la historia de la marina castellana y de los establecimientos espanoles en India, coordinada e ilustrada por Don Martin Fernandez de Navarrete. Prologo de J. Natalicio Gonzalez.* 5 vols. Buenos Aires: Editorial Guarania, 1945.

Nunes, Pedro. *Tratado en defensam da carta de marear.* Transcribed and edited by Francisco Maria Esteves Pereira. In *Revista de Engenharia Militar.* Lisbon: Typ. do Commercio, 1911–12.

Oviedo, Gonzalo Fernández de. *Historia general de las Indias.* Seville: Juan Cromberger, 1535.

———. *Historia general y natural de las Indias.* 4 vols. Madrid: Imprenta de la Real Academia de la Historia, 1851–55.

Padrón, Francisco Morales. *Primeras cartas sobre America (1493–1503).* Seville: Secretariado de Publicaciones Universidad de Sevilla, 1990.

Pereira, Duarte Pacheco. *Esmeraldo de Situ Orbis.* Lisbon: Imprensa Nacional, 1892.

Peres, Damião. *Os mais antigos roteiros da Guiné.* Lisbon, 1952.

Plato. *Timaeus.* Translated by Donald J. Zeyl. Indianapolis: Hackett, 2000.

Pliny, the Elder. *Natural History.* Translated by H. Rackham. Cambridge, MA: Harvard University Press, 2014.

Polo, Marco, Philippe Ménard, Marie-Luce Chênerie, and Michèle Guéret-Laferté. *Le Devisement du Monde. 6 vols.* Geneva: Droz, 2001.

Portugal. Tratados, etc., 1494. "Minuta do Tratado de Tordesilhas," n.d. Mss-5-25. Biblioteca Nacional de Portugal, Lisbon.

Prottengeier, Alvin E., and John Parker. *From Lisbon to Calicut.* Minneapolis: University of Minnesota Press, 1956.

Ramusio, Giovanni Battista. *Navigazioni e Viaggi.* 6 vols. Torino: Einaudi, 1978.

Sanuto, Marino, Rinaldo Fulin, and Deputazione di storia patria per le Venezie. *I diarii di Marino Sanuto.* 58 vols. Bologna: Forni, 1969.

Sanz, Carlos. *Bibliotheca Americana Vetustissima: últimas adiciones.* Vol. 1. Madrid: Librería General V. Suarez, 1960.

Schedel, Hartmann. *Das Buch der Croniken und Geschichten.* Nuremberg: Anton Koberger, 1493.

———. *Liber Chronicarum.* Nuremberg, 1493.

———. *The Nuremberg Chronicle: A Facsimile of Hartmann Schedel's Buch der Chroniken, Printed by Anton Koberger in 1493.* New York: Landmark Press, 1979.

Seville, Isidore de. *Etymologiae* Augsburg: Günther Zainer, 1472. Bayerische Staatsbibliothek, Munich.

———. *The Etymologies of Isidore of Seville.* Edited by Stephen A. Barney, W. J. Lewis, J. A. Beach, and Oliver Berghof. Cambridge: Cambridge University Press, 2006.

Staden, Hans, Neil L. Whitehead, and Michael Harbsmeier. *Hans Staden's True History: An Account of Cannibal Captivity in Brazil.* Durham: Duke University Press, 2008.

Syllacius, Nicolaus. *De Insulis Meridiani atque Indici Maris Nuper Inventis. With a Translation by John Mulligan.* New York: 1859.

Trevisan, Angelo, and Angela Caracciolo Aricò. *Lettere sul Nuovo Mondo: Granada 1501.* Venice: Albrizzi Editore, 1993.

Trithemius, Johannes. *In Praise of Scribes. De Laude Scriptorum.* Edited by Klaus Arnold; translated by Roland Behrendt. Lawrence, KS: Coronado Press, 1974.

———. *Opera historica.* Frankfurt: Typis Wechelianis apud Claudium, 1601.

Vespucci, Amerigo. *The Letters of Amerigo Vespucci and other Documents Illustrative of his Career.* Edited by Clements Markham. Hakluyt Society No. 90. London: Chas. J. Clark, 1894.

———. *Mundus Novus.* Augsburg: Johannes Otmar, 1504. John Carter Brown Library of Brown University, Providence, RI.

Vespucci, Amerigo, and Luciano Formisano. *Lettere di Viaggio.* Milano: A. Mondadori, 1985.

———. *Letters from a New World: Amerigo Vespucci's Discovery of America.* Translated by David Jacobson. New York: Marsilio, 1992.

Waldseemüller, Martin. *Cosmographiae Introductio: cum quibusdam geometriae ac astronomiae principiis ad eam rem necessariis. Insuper Quattuor Americi Vespucij nauigationes: Vniuersalis cosmographiæ descriptio tam in solido [quam] plano, eis etiam insertis quæ Ptholom[a]eo ignota a nuperis reperta sunt.* St. Dié: Walter et Nicolaus Lud, 1507. John Carter Brown Library at Brown University, Providence, RI. https://archive.org/details/cosmographiaeintoowald_0.

Waldseemüller, Martin, Joseph Fischer, and Franz Wieser. *Die Älteste Karte Mit Dem Namen Amerika Aus Dem Jahre 1507 Und Die Carta Marina Aus Dem Jahre 1516 Des M. Waldseemüller (Ilacomilus).* Translated by George Jacob Pickel. Innsbruck and London: Wagner and H. Stevens, Son, & Stiles, 1903.

Waldseemüller, Martin, Charles George Herbermann, Joseph Fischer, and Franz Wieser. *The Cosmographiæ Introductio of Martin Waldseemüller in Facsimile, Followed by the Four Voyages of Amerigo Vespucci, with their Translation into English . . .* New York: United States Catholic Historical Society, 1907.

Waldseemüller, Martin, and John W. Hessler. *The Naming of America: Martin Waldseemüller's 1507 World Map and the Cosmographiae Introductio.* London: Giles, 2008.

Zurara, Gomes Eanes de. *Chronica do descobrimento e conquista de Guiné.* Paris: J. P. Aillaud, 1841.

———. *Crónica dos feitos notáveis que se passaram na conquista da Guiné por mandado do Infante D. Henrique.* Lisbon: Academia Portuguesa da História, 1978.

———. *The Discovery and Conquest of Guinea.* Translated by Charles Raymond Beazley and Edgar Prestage. Hakluyt Society, no. XCV. New York: Burt Franklin, 1896.

Modern Sources

Abulafia, David. *The Discovery of Mankind: Atlantic Encounters in the Age of Columbus.* New Haven: Yale University Press, 2008.

Alberti, Leon Battista 1404–1472. *Leon Battista Alberti's Delineation of the City of Rome (Descriptio Vrbis Romæ).* Tempe: Arizona Center for Medieval and Renaissance Studies, 2007.

Alberti, Leon Battista, and Rocco Sinisgalli. *Leon Battista Alberti: On Painting; A New Translation and Critical Edition.* New York: Cambridge University Press, 2011.

Albuquerque, Luís de. *A Comissão de cartografia e a cartografia portuguesa antiga.* Lisbon: Instituto de Investigação Científica Tropical, 1984.

———. *Historia de la navegación portuguesa.* Lisbon: Editorial MAPFRE, 1991.

Albuquerque, Luís de, and Francisco Contente Domingues. *Dicionário de história dos descobrimentos portugueses.* 2 vols. Lisbon: Caminho, 1994.

Albuquerque, Luís de, and J. Lopes Tavares. "Algumas observações sobre o planisfério 'Cantino' (1502)." *Revista do Centro de Estudos Geográficos* 3, nos. 22–23 (1967).

Alchon, Suzanne Austin. *A Pest in the Land: New World Epidemics in a Global Perspective.* Albuquerque: University of New Mexico Press, 2003.

Alden, John E. and Dennis C. Landis. *European Americana: A Chronological Guide to Works Printed in Europe Relating to the Americas, 1493–1776.* Vol. 1. New York: Readex Books, 1980.

Alegria, Maria Fernanda, Suzanne Daveau, João Carlos Garcia, and Francesc Relaño. "Portuguese Cartography in the Renaissance." In *The History of Cartography*, 3: *Cartography in the European Renaissance*, edited by David Woodward, 975–1068. Chicago: University of Chicago Press, 2007.

Alegria, Maria Fernanda, Suzanne Daveau, and Francesc Relaño. *História da cartografia portuguesa: séculos XV a XVII.* Porto, Portugal: Fio da Palavra, 2012.

Alegria, Maria Fernanda, João Carlos Garcia, and Francesc Relaño. "Cartografia e viagens." In *História da expansão portuguesa*, vol. 1: *A formação do império (1415–1570)*, edited by Francisco Bethencourt and K. N. Chaudhuri, 26–61. Lisbon: Temas e Debates, 1998.

Almagià, Roberto. "On the Cartographic Work of Francesco Rosselli." *Imago Mundi* 8 (1951): 27–34.

Alpers, Svetlana. "The Mapping Impulse in Dutch Art." In *Art and Cartography: Six Historical Essays*, edited by David Woodward, 51–96. Chicago: University of Chicago Press, 1987.

Amaral, Joaquim Ferreira do. *Pedro Reinel me fez: A volta de um mapa dos descobrimentos*. Lisbon: Quetzal Editores, 1995.

Andrien, Kenneth J., and Peter Hulme. "Colonial Encounters: Europe and the Native Caribbean, 1492–1787." *Sixteenth Century Journal* 24 (1993): 922–23.

Arciniegas, Germán. *Amerigo and the New World: The Life and Times of Amerigo Vespucci*. Translated by Harriet de Onís. New York: Knopf, 1955.

Arens, W. *The Man-Eating Myth: Anthropology and Anthropophagy*. New York: Oxford University Press, 1979.

Armitage, David, Alison Bashford, and Sujit Sivasundaram, eds. *Oceanic Histories*. Cambridge: Cambridge University Press, 2018.

Armitage, David, and Michael Braddick. *The British Atlantic World, 1500–1800*. Basingstoke, UK: Palgrave Macmillan, 2002.

Arnaud, Pascal. "Ancient Mariners between Experience and 'Common Sense' Geography." In *Features of Common Sense Geography: Implicit Knowledge Structures in Ancient Geographical Texts*, edited by Klaus Gues and Martin Thiering, 39–68. Berlin: Antike Kulture und Geschichte, 2014.

Arnold, A. James, Julio Rodríguez-Luis, and J. Michael Dash. *A History of Literature in the Caribbean*. Amsterdam: J. Benjamins, 1994.

Astengo, Corradino. "The Renaissance Chart Tradition in the Mediterranean." In *The History of Cartography*, vol. 3: *Cartography in the European Renaissance*, edited by David Woodward, 174–262. Chicago: University of Chicago Press, 2007.

Aujac, Germaine, and the editors. "The Foundations of Theoretical Cartography in Archaic and Classical Greece." In *The History of Cartography*, vol. 1: *Cartography in Prehistoric, Ancient, and Medieval Europe and the Mediterranean*, edited by J. B. Harley and David Woodward, 130–47. Chicago: University of Chicago Press, 1987.

———. "The Growth of an Empirical Cartography in Hellenistic Greece." In *The History of Cartography*, vol. 1: *Cartography in Prehistoric, Ancient, and Medieval Europe and the Mediterranean*, edited by J. B. Harley and David Woodward, 148–60. Chicago: University of Chicago Press, 1989.

"Avviso Agli Utenti—Chiusura precauzionale di alcuni locali e fondi esclusi dalla consultazione." Archivio di Stato Modena, 24 April 2012. http://www.asmo.beniculturali .it/index.php?it/210/archivio-news/31/avviso-agli-utenti-chiusura-precauzionale-di -alcuni-locali-e-fondi-esclusi-dalla-consultazione.

Bagrow, Leo, and R. A. Skelton, *History of Cartography*. 2nd ed. Chicago: Precedent, 1985.

Bailyn, Bernard. *Atlantic History: Concept and Contours*. Cambridge, MA: Harvard University Press, 2005.

Ballon, Hilary, and David Friedman. "Portraying the City in Early Modern Europe: Measurement, Representation, and Planning." In *The History of Cartography*, vol. 3: *Cartography in the European Renaissance*, edited by David Woodward, 680–704. Chicago: University of Chicago Press, 2007.

Barber, Peter. "Context Is Everything . . ." Keynote Address presented at the International Conference on the History of Cartography, Amsterdam, 14 July 2019.

Barlow, Tani. "Advertising Ephemera and the Angel of History." *Positions: asia critique* 20 (2012): 111–58.

Barr, Juliana. "Beyond the 'Atlantic World': Early American History as Viewed from the West." *OAH Magazine of History* 25 (2011): 13–18.

Bartrum, Giulia. *German Renaissance Prints, 1490–1550.* London: British Museum Press, 1995.

Bayerische Staatsbibliothek, Munich. "Bavarian State Library Confirms That Its Own 'Waldseemüller' Globe Gores Map Is a Forgery." Press Statement, 2 February 2018. https://www.bsb-muenchen.de/en/article/bavarian-state-library-confirms-that-its-own -waldseemueller-globe-gores-map-is-a-forgery-2280/.

Baynton-Williams, Ashley, and Miles Baynton-Williams. *New Worlds: Maps from the Age of Discovery.* London: Quercus, 2006.

Bedini, Silvio A. *The Pope's Elephant.* Manchester: Carcanet Press, 1997.

Benjamin, Thomas. *The Atlantic World: Europeans, Africans, Indians and Their Shared History, 1400–1900.* New York: Cambridge University Press, 2009.

Berenbaum, Michael, and Fred Skolnik, eds. "Gog and Magog." In *Encyclopaedia Judaica,* 7:683–84. Detroit: Macmillan Reference USA, 2007.

Berggren, J. Lennart, Alexander Jones, and Ptolemy. *Ptolemy's Geography: An Annotated Translation of the Theoretical Chapters.* Princeton: Princeton University Press, 2000.

Bethencourt, Francisco, and K. N. Chaudhuri. *História da expansão portuguesa.* Vol. 1: *A Formação do Império (1415–1570).* Lisbon: Temas e Debates, 1998.

Bethencourt, Francisco, and Diogo Ramada Curto. *Portuguese Oceanic Expansion, 1400–1800.* Cambridge: Cambridge University Press, 2007.

Betz, Richard L. *The Mapping of Africa: A Cartobibliography of Printed Maps of the African Continent to 1700.* Utrecht: Hes & de Graaf, 2007.

Billion, Philipp. "A Newly Discovered Chart Fragment from the Lucca Archives, Italy." *Imago Mundi* 63 (11 January 2011): 1–21.

Bindman, David, and Henry Louis Gates, Jr. *The Image of the Black in Western Art.* 5 Vols. Cambridge, MA: Harvard University Press, 2010.

Bini, Mauro. "Nota storico-codicologica." In *L'atlante di Borso d'Este: la cosmographia di Claudio Tolomeo della Biblioteca Estense universitaria di Modena,* 33–35. Modena: Il Bulino, 2006.

Blanding, Michael. "Why Experts Don't Believe This Is a Rare First Map of America." *New York Times,* 11 December 2017, sec. Arts.

Bleichmar, Daniela, Paula De Vos, Kristin Huffine, and Kevin Sheehan, eds. *Science in the Spanish and Portuguese Empires, 1500–1800.* Stanford: Stanford University Press, 2009.

Boehrer, Bruce Thomas. *Parrot Culture: Our 2,500-Year-Long Fascination with the World's Most Talkative Bird.* Philadelphia: University of Pennsylvania Press, 2004.

Boelhower, William. "Framing Anew Ocean Genealogy: The Case of Venetian Cartography in the Early Modern Period." *Atlantic Studies,* 13 March 2018.

———. "Inventing America: The Culture of the Map." *Revue Française d'études Américaines,* no. 36 (1988): 211–24.

———. "The Rise of the New Atlantic Studies Matrix." *American Literary History* 20, nos. 1–2 (2008): 83–101.

Bouloux, Nathalie. "L'Insularium Illustratum d'Henricus Martellus." *Historical Review/La Revue Historique* 9 (2012): 77–94.

Brooks, George E. *Landlords and Strangers: Ecology, Society, and Trade in Western Africa, 1000–1630.* Boulder, CO: Westview Press, 1993.

Brotton, Jerry. *A History of the World in Twelve Maps.* New York: Penguin Books, 2012.

———. *Trading Territories: Mapping the Early Modern World.* Ithaca: Cornell University Press, 1998.

Brückner, Martin. *Early American Cartographies.* Chapel Hill: University of North Carolina Press, 2011.

———. *The Social Life of Maps in America, 1750–1860.* Chapel Hill: University of North Carolina Press, 2017.

Buisseret, David, ed. *From Sea Charts to Satellite Images: Interpreting North American History through Maps.* Chicago: University of Chicago Press, 1990.

———. *The Mapmakers' Quest: Depicting New Worlds in Renaissance Europe.* Oxford: Oxford University Press, 2003.

———, ed. *Monarchs, Ministers, and Maps: The Emergence of Cartography as a Tool of Government in Early Modern Europe.* Chicago: University of Chicago Press, 1992.

———. *Tools of Empire: Ships and Maps in the Process of Westward Expansion.* Chicago: Newberry Library, 1986.

Burden, Philip D. *The Mapping of North America: A List of Printed Maps.* Vol. 1: *1511–1670.* Rickmansworth, UK: Raleigh, 1996.

Campbell, Stephen J. *Cosmè Tura of Ferrara: Style, Politics, and the Renaissance City, 1450–1495.* New Haven: Yale University Press, 1997.

Campbell, Tony. "Census of Pre-Sixteenth-Century Portolan Charts." *Imago Mundi* 38 (1986): 67–94.

———. "Maphistory / History of Cartography, Workshops." http://www.maphistory.info /PortolanAttributions.html.

———. "Portolan Charts from the Late Thirteenth Century to 1500." In *The History of Cartography*, vol. 1: *Cartography in Prehistoric, Ancient, Medieval Europe and the Mediterranean*, edited by J. B. Harley and David Woodward, 371–463. Chicago: University of Chicago Press, 1987.

Campbell, Tony, and Marcel Destombes. *The Earliest Printed Maps, 1472–1500.* Berkeley: University of California Press, 1987.

Cañizares-Esguerra, Jorge. *Entangled Empires: The Anglo-Iberian Atlantic, 1500–1830.* Philadelphia: University of Pennsylvania Press, 2018.

———. *Puritan Conquistadors: Iberianizing the Atlantic, 1550–1700.* Stanford: Stanford University Press, 2006.

Canny, Nicholas, and Philip Morgan. *The Oxford Handbook of the Atlantic World: 1450–1850.* Oxford: Oxford University Press, 2011.

Canny, Nicholas, and Anthony Pagden, eds. *Colonial Identity in the Atlantic World, 1500–1800.* Princeton: Princeton University Press, 1989.

Caraci, Giuseppe. "An Unknown Nautical Chart of Grazioso Benincasa, 1468." *Imago Mundi* 7 (1950): 18–31.

Carlton, Genevieve. *Worldly Consumers: The Demand for Maps in Renaissance Italy.* Chicago: University of Chicago Press, 2015.

Carney, Judith Ann, and Richard Nicholas Rosomoff. *In the Shadow of Slavery: Africa's Botanical Legacy in the Atlantic World.* Berkeley: University of California Press, 2009.

Carter, Harry Graham. *A View of Early Typography up to about 1600.* Oxford: Clarendon Press, 1969.

Castro, Eduardo Batalha Viveiros de. *From the Enemy's Point of View: Humanity and Divinity in an Amazonian Society,* Translated by Catherine V. Howard. Chicago: University of Chicago Press, 1992.

Cattaneo, Angelo. "Découvertes littéraires et géographiques au xve siècle. Le 'Portolano 1' de la Bibliothèque nationale centrale de Florence." *Médiévales. Langues, Textes, Histoire,* no. 58 (2010): 79–98.

———. *Fra Mauro's Mappa Mundi and Fifteenth-Century Venice.* Turnhout: Brepols, 2011.

Ceva, Juan. *The Cresques Project.* http://www.cresquesproject.net/home.

Charles, John. *Allies at Odds: The Andean Church and Its Indigenous Agents, 1583–1671.* Albuquerque: University of New Mexico Press, 2010.

Chiappelli, Fredi, with Michael J. B Allen and Robert Louis Benson, eds. *First Images of America: The Impact of the New World on the Old.* 2 vols. Berkeley: University of California Press, 1976.

Chiappini, Luciano. "1598: Chronicle of a Changement." *Ferrara: Voci di una città* 7, no. 6 (1997). https://rivista.fondazionecarife.it/en/1997/item/383-1598-diario-di-una-svolta.

Coates, Timothy. Review of *Atlantic History* by Bernard Bailyn. *e-journal of Portuguese History* 3 (2005). https://www.brown.edu/Departments/Portuguese_Brazilian_Studies/ejph/html/Summer05.html.

Coclanis, Peter A. Review of *Atlantic History* by Bernard Bailyn. *Business History Review* 80 (2006): 171–73.

Coffman, D'Maris, Adrian Leonard, and William O'Reilly, Eds. *The Atlantic World.* London: Routledge, 2014.

Colin, Susi. "Woodcutters and Cannibals: Brazilian Indians as Seen on Early Maps." In *America: Early Maps of the New World,* edited by Hans Wolff, 175–81. Munich: Prestel, 1992.

Collins, Edward. "Francisco Faleiro and Scientific Methodology at the Casa de La Contratación in the Sixteenth Century." *Imago Mundi* 65 (2013): 25–36.

Connolly, Daniel K., and Matthew Paris. *The Maps of Matthew Paris: Medieval Journeys through Space, Time and Liturgy.* Woodbridge, NY: Boydell Press, 2009.

Conway, Stephen. Review of *Atlantic History* by Bernard Bailyn. *History* 92 (2007): 97–98.

Cook, Noble David. *Born to Die: Disease and New World Conquest, 1492–1650.* Cambridge: Cambridge University Press, 1998.

———. *Demographic Collapse: Indian Peru, 1520–1620.* Cambridge: Cambridge University Press, 1982.

Cook, Noble David, and W. George Lovell. *Secret Judgments of God: Old World Disease in Colonial Spanish America.* Norman: University of Oklahoma Press, 2001.

Cook, Sherburne F., and Woodrow Borah. *The Indian Population of Central Mexico, 1531–1610.* Berkeley: University of California Press, 1960.

Cortesão, Armando. "A Hitherto Unrecognized Map by Pedro Reinel in the British Museum." *Geographical Journal* 87 (1936): 518–24.

———. "A Carta Náutica de 1424." In *Esparsos,* vol. 3. Coimbra: Acta Universitatis Conimbrigensis, 1975.

———. *Cartografia e cartógrafos portugueses dos séculos XV e XVI. (Contribuição para um estudo completo).* Lisbon: Edição da "Seara nova," 1935.

————. *Cartografia portuguesa antiga*. Lisbon: Comissão Executiva das Comemorações do Quinto Centenário da Morte do Infante D. Henrique, 1960.

————. *História da cartografia portuguesa*. 2 vols. Lisbon: Junta de Investigações de Ultramar, 1969.

————. *O Problema da origem da carta portulano*. Coimbra: Junta de Investigações do Ultramar, 1966.

————. *Os Homens (cartógrafos portugueses do século XVI)*. Coimbra: Imprensa da Universidade, 1932.

————. *Subsídios para a história do descobrimento da Guiné e Cabo Verde*. Coimbra: Imprensa da Universidade, 1932.

Cortesão, Armando, and Avelino Teixeira Mota. *Portugaliae Monumenta Cartographica*. 6 vols. Lisbon: Comemorações do V Centenário da Morte do Infante D. Henrique, 1960.

————. *Tabularum Geographicarum Lusitanorum Specimen*. Lisbon: Comissão Ultramarina das Comemorações do V Centenario da Morte do Infante D. Henrique, 1960.

Cortesão, Jaime. *A colonização do Brasil*. 3 vols. Lisbon: Portugalia editora, 1969.

————. *Os descobrimentos pré-colombinos dos portugueses*. Lisbon: Portugália Editora, 1966.

————. *Os descobrimentos portugueses*. 3rd ed. 6 vols. Lisbon: Livros Horizonte, 1975.

Cosgrove, Denis E. "Images of Renaissance Cosmography, 1450–1650." In *The History of Cartography*, vol. 3: *Cartography in the European Renaissance*, edited by David Woodward, 55–98. Chicago: Chicago University Press, 2007.

Cosner, Charlotte. *The Golden Leaf: How Tobacco Shaped Cuba and the Atlantic World*. Nashville: Vanderbilt University Press, 2015.

Couto, Jorge. *A construção do Brasil: Ameríndios, Portugueses e Africanos, do início do povoamento a finais de quinhentos*. Lisbon: Edições Cosmos, 1997.

Crone, G. R. *Maps and Their Makers: An Introduction to the History of Cartography*. London: Hutchinson's University Library, 1953.

————. "The Origin of the Name Antillia." *The Geographical Journal* 91, no. 3 (1938): 260–62.

Crosby, Alfred W. *The Columbian Exchange: Biological and Cultural Consequences of 1492*. 30th anniversary edition. Westport: Praeger, 2003.

————. *Ecological Imperialism: The Biological Expansion of Europe, 900–1900*. 2nd ed. Cambridge: Cambridge University Press, 2004.

Crowe, Jonathan. "More on the Waldseemüller Globe Gores Auction." Map Room, 28 November 2017, http://www.maproomblog.com/2017/11/more-on-the-waldseemuller-globe-gores-auction/.

Dackerman, Susan, ed. *Prints and the Pursuit of Knowledge in Early Modern Europe*. Cambridge, MA: Harvard Art Museums, 2011.

Dalché, Patrick Gautier. "The Reception of Ptolemy's Geography (End of the Fourteenth to Beginning of the Sixteenth Century)." In *The History of Cartography*, vol. 3: *Cartography in the European Renaissance*, edited by David Woodward, 285–364. Chicago: University of Chicago Press, 2007.

Daniels, Christine, and Michael V. Kennedy. *Negotiated Empires: Centers and Peripheries in the Americas, 1500–1820*. New York: Routledge, 2002.

Dann, John C. "Ephemera Collecting—A Growing Field, Hard to Define." Ephemera Society. http://www.ephemerasociety.org/blog/?p=260.

Daston, Lorraine, and Katharine Park. *Wonders and the Order of Nature, 1150–1750*. New York: Zone Books, 1998.

Davies, Arthur. "Behaim, Martellus and Columbus." *Geographical Journal* 143 (1977): 451–59.

Davies, Surekha. "America and Amerindians in Sebastian Münster's *Cosmographiae Universalis Libri VI* (1550)." *Renaissance Studies* 25 (June 2011): 351–73.

———. "Depictions of Brazilians on French Maps, 1542–1555." *Historical Journal* 55 (2012): 317–48.

———. *Renaissance Ethnography and the Invention of the Human: New Worlds, Maps and Monsters*. Cambridge: Cambridge University Press, 2016.

———. "The Navigational Iconography of Diogo Ribeiro's 1529 Vatican Planisphere." *Imago Mundi* 55 (October 2003): 103–13.

Dean, Warren. *With Broadax and Firebrand: The Destruction of the Brazilian Atlantic Coastal Forest*. Berkeley: University of California Press, 1997.

Denucé, Jean. *Inventaire des Affaitadi, banquiers italiens à Anvers, de l'année 1568*. Antwerp: Éditions de Sikkel, 1937.

———. *L'Afrique au XVIe siècle et le commerce anversois*. Anvers: De Sikkel, 1937.

———. *Les origines de la cartographie portugaise et les cartes des Reinel*. Amsterdam: Meridian, 1963.

Devise, Jean, and Michel Mollat. *From the Early Christian Era to the Age of Discovery: The Image of the Black in Western Art*, vol. 2, pt. 2. Edited by David Bindman and Henry Louis Gates Jr. Cambridge, MA: Harvard University Press, 2010.

Diamond, Jared M. *Guns, Germs, and Steel: The Fates of Human Societies*. New York: W. W. Norton, 2005.

Dias, Carlos Malheiros, Alfredo Roque Gameiro, and Ernesto Julio de Carvalho e Vasconcel-los. *História da colonização portuguesa do Brasil: edição monumental comemorativa do primeiro centenário da independência do Brasil*. 3 vols. Porto: Lithografia Nacional, 1921.

Díaz y Díaz, Manuel C. "Isidore of Seville." In *Encyclopedia of Religion*, edited by Lindsay Jones, 7:4556–57. 2nd ed. Detroit: Macmillan Reference USA, 2005.

Dickason, Olive P. *The Myth of the Savage and the Beginnings of French Colonialism in the Americas*. Edmonton, Alberta: University of Alberta Press, 1997.

Dickinson, Helen A. *German Masters of Art*. New York: Frederick A. Stokes, 1914.

Dilke, O. A. W. "Cartography in the Ancient World: An Introduction." In *The History of Cartography*, vol. 1: *Cartography in the Prehistoric, Ancient, and Medieval Europe and the Mediterranean*, edited by J. B. Harley and David Woodward, 105–6. Chicago: University of Chicago Press, 1987.

———. "Cartography in the Byzantine Empire." In *The History of Cartography*, vol. 1: *Cartography in the Prehistoric, Ancient, and Medieval Europe and the Mediterranean*, edited by J. B. Harley and David Woodward, 258–75. Chicago: University of Chicago Press, 1987.

———. "The Culmination of Greek Cartography in Ptolemy." In *The History of Cartography*, vol. 1: *Cartography in Ancient Europe and the Mediterranean*, edited by J. B. Harley and David Woodward, 177–200. Chicago: Chicago University Press, 1987.

———. *Greek and Roman Maps*. London: Thames and Hudson, 1985.

―――. "Itineraries and Geographical Maps in the Early and Late Roman Empires." In *The History of Cartography*, vol. 1: *Cartography in the Prehistoric, Ancient, and Medieval Europe and the Mediterranean*, edited by J. B. Harley and David Woodward, 234–57. Chicago: University of Chicago Press, 1987.

Domingues, Francisco Contente. "Colombo e a política de siglio na historiografia portuguesa." *Revista Mare Liberum* 1 (1990): 105–16.

Drisin, Adam. "Intricate Fictions: Mapping Princely Authority in a Sixteenth-Century Florentine Urban Plan." *Journal of Architectural Education* 57, no. 4 (2004): 41–55.

Duffy, Eve M., and Alida C. Metcalf. *The Return of Hans Staden: A Go-between in the Atlantic World*. Baltimore: Johns Hopkins University Press, 2012.

Dunkelman, Arthur. *The Jay I. Kislak Collection at the Library of Congress*. Washington, DC: Library of Congress, 2007.

Dursteler, Eric. "Reverberations of the Voyages of Discovery in Venice, ca. 1501: The Trevisan Manuscript in the Library of Congress." *Mediterranean Studies* 9 (2000): 43–64.

Dutschke, C. W., et al. *Guide to Medieval and Renaissance Manuscripts in the Huntington Library*. 2 vols. San Marino, CA: The Library, 1989.

Duzer, Chet Van, and Sandra Sáenz-López Pérez. "Tres Filii Noe Diviserunt Orbem Post Diluvium: The World Map in British Library Add. MS 37049." *Word and Image* 26 (2009): 21–39.

Dym, Jordana, and Karl Offen. *Mapping Latin America: A Cartographic Reader*. Chicago: University of Chicago Press, 2011.

Eames, Wilberforce. *Description of a Wood Engraving Illustrating the South American Indians (1505)*. New York: New York Public Library, 1922.

Ebert, Christopher. "European Competition and Cooperation in Pre-modern Globalization: 'Portuguese' West and Central Africa. 1500–1600." *African Economic History* 36 (2008): 53–78.

Edgerton, Douglas R., Alison Games, Jane G. Landers, Kris Lane, and Donald R. Wright. *The Atlantic World: 1440–1888*. Wheeling, IL: Harlan Davidson, 2007.

Edgerton, Samuel Y. "From Mental Matrix to *Mappamundi* to Christian Empire: The Heritage of Ptolemaic Cartography in the Renaissance." In *Art and Cartography: Six Historical Essays*, edited by David Woodward, 10–50. Chicago: University of Chicago Press, 1987.

Edney, Matthew H. "The Irony of Imperial Mapping." In *The Imperial Map: Cartography and the Mastery of Empire*, edited by James Akerman, 11–45. Chicago: University of Chicago Press, 2009.

―――. "Knowledge and Cartography in the Early Atlantic." In *The Oxford Handbook of the Atlantic World, 1450–1850*, edited by Nicolas Canny and Philip Morgan, 87–112. Oxford: Oxford University Press, 2011.

―――. *Mapping an Empire: The Geographical Construction of British India, 1765–1843*. Chicago: University of Chicago Press, 1997.

Edson, Evelyn. *The World Map, 1300–1492: The Persistence of Tradition and Transformation*. Baltimore: Johns Hopkins University Press, 2007.

E.H., "A New View of the Vespucci Problem." *Geographical Journal* 66 (1925): 339–44.

Eisler, Colin T. *Dürer's Animals*. Washington, DC. Smithsonian Institution Press, 1991.

Elbl, Ivana. "The Volume of the Early Atlantic Slave Trade, 1450–1521." *Journal of African History* 38 (1997): 31–75.

Elliott, J. H. *Empires of the Atlantic World: Britain and Spain in America, 1492–1830.* New Haven: Yale University Press, 2006.

Evans, James. *The History and Practice of Ancient Astronomy.* New York: Oxford University Press, 1998.

Falchetta, Piero. *Fra Mauro's World Map: With a Commentary and Translations of the Inscriptions.* Turnhout and Venice: Brepols & Biblioteca Nazionale Marciana, 2006.

Falola, Toyin, and Matt D. Childs. *The Yoruba Diaspora in the Atlantic World.* Bloomington: Indiana University Press, 2005.

Federzoni, Laura. "La geografia di Tolomeo." In *L'atlante di Borso d'Este: la cosmographia di Claudio Tolomeo della Biblioteca Estense universitaria di Modena*, 11–31. Modena: Il Bulino, 2006.

Fernandes, Fernando Lourenço. *O Planisfério de Cantino e o Brasil: uma introdução à cartologia política dos descobrimentos e o Atlântico Sul.* Lisbon: Academia de Marinha, 2003.

Fernandes, Florestan. *Organização social dos Tupinambá.* 2nd ed. São Paulo: Difusão Européia do Livro, 1963.

Fernández-Armesto, Felipe. *Amerigo: The Man Who Gave His Name to America.* New York: Random House, 2007.

———. *Before Columbus: Exploration and Colonisation from the Mediterranean to the Atlantic, 1229–1492.* Basingstoke, UK: Macmillan Education, 1987.

———. "Maps and Exploration in the Sixteenth and Early Seventeenth Centuries." In *The History of Cartography*, vol. 3: *Cartography in the European Renaissance*, edited by David Woodward, 738–58. Chicago: University of Chicago Press, 2007.

Ferrar, Michael. "Portolan Charts: Construction and Copying [Occam's Razor Methodology (Lex Parsimoniae)]." Paper. *Cartography Unchained*, https://www.cartographyunchained.com/chp01/.

Ferreira, Roquinaldo Amaral. *Cross-Cultural Exchange in the Atlantic World: Angola and Brazil during the Era of the Slave Trade.* New York: Cambridge University Press, 2012.

Findlen, Paula. "Possessing the Past: The Material World of the Italian Renaissance." *American Historical Review* 103 (1998): 83–114.

Fiorani, Francesca. "Cycles of Painted Maps in the Renaissance." In *The History of Cartography*, vol. 3: *Cartography in the European Renaissance*, edited by David Woodward, 804–30. Chicago: University of Chicago Press, 2007.

———. *The Marvel of Maps: Art, Cartography and Politics in Renaissance Italy.* New Haven: Yale University Press, 2005.

Fynn-Paul, Jeff, and Damian Alan Pargas, eds. *Slaving Zones: Cultural Identities, Ideologies, and Institutions in the Evolution of Global Slavery.* Leiden: Brill, 2018.

Foncin, Myriem, Marcel Destombes, and Monique de la Roncière. *Catalogue des cartes nautiques sur vélin conservées au Département des cartes et plans.* Paris: Bibliothèque nationale, 1963.

Forsyth, Donald W. "Three Cheers for Hans Staden: The Case for Brazilian Cannibalism." *Ethnohistory* 32 (1985): 17–36.

"Foreign Bodies." In *Connecting Collections at Manchester and Melbourne*, https://connectingcollections-manmel.com/.

Games, Alison. "Atlantic History: Definitions, Challenges, and Opportunities." *American Historical Review* III (2006): 741–57.

———. Review of *Atlantic History* by Bernard Bailyn. *American Historical Review* III(2006): 434–35.

———. "Teaching Atlantic History." *Itinerario* 23, no. 2 (July 1999): 162–74.

Garfield, Robert. *A History of São Tome Island, 1470–1655: The Key to Guinea*. San Francisco: Edwin Mellen Press, 1992.

Garrigus, John D., and Christopher Morris, eds. *Assumed Identities: The Meanings of Race in the Atlantic World*. College Station: Texas A&M University Press, 2010.

Gaspar, Joaquim Alves. "Blunders, Errors and Entanglements: Scrutinizing the Cantino Planisphere with a Cartometric Eye." *Imago Mundi* 64 (2012): 181–200.

———. "From the Portolan Chart of the Mediterranean to the Latitude Chart of the Atlantic: Cartometric Analysis and Modeling." PhD diss., Universidade Nova de Lisboa, 2010.

———. "The Myth of the Square Chart." *e-Perimetron* 2 (2007): 66–79.

———. "Revisitando o planisfério de Juan de La Cosa: Uma Abordagem Cartométrica." Biblioteca Nacional de Portugal, Lisbon, Portugal: Academia.edu, 2012.

———. "Using Empirical Map Projections for Modeling Early Nautical Charts." *Advances in Cartography and GIScience* 2 (2011): 227–47.

Gaspar, Joaquim Alves, and Henrique Leitão. "Squaring the Circle: How Mercator Constructed His Projection in 1569." *Imago Mundi* 66 (2014): 1–24.

Gelderblom, Oscar C. "From Antwerp to Amsterdam: The Contribution of Merchants from the Southern Netherlands to the Commercial Expansion of Amsterdam (c. 1540–1609)." *Review (Fernand Braudel Center)* 26 (2003): 247–82.

George, Wilma B. *Animals and Maps*. Berkeley: University of California Press, 1969.

Gerbi, Antonello. *Nature in the New World: From Christopher Columbus to Gonzalo Fernández de Oviedo*. Pittsburgh: University of Pittsburgh Press, 2010.

Gerbner, Katharine. *Christian Slavery: Conversion and Race in the Protestant Atlantic World*. Philadelphia: University of Pennsylvania Press, 2019.

Gier, Helmut, and Johannes Janota. *Augsburger Buchdruck und Verlagswesen: Von den Anfängen bis zur Gegenwar*. Wiesbaden: Harrassowitz, 1997.

Gill, Christopher. *Plato's Atlantis Story: Text, Translation and Commentary*. Liverpool: Liverpool University Press, 2017.

Goerz, Guenther, and Martin Scholz. "Semantic Annotation for Medieval Cartography: The Example of the Behaim Globe of 1492." *e-Perimetron* 8 (2013): 8–20.

Goñi-Gaztambide, José. "Bernardino López de Carvajal y las bulas alejandrinas." *Anuario de Historia de la Iglesia*, 1 (1992): 93–112.

Gould, Eliga H. *Among the Powers of the Earth: The American Revolution and the Making of a New World Empire*. Reprint edition. Cambridge, MA: Harvard University Press, 2014.

Gow, Andrew. "Gog and Magog on Mappaemundi and Early Printed World Maps: Orientalizing Ethnography in the Apocalyptic Tradition." *Journal of Early Modern History* 2 (1998): 61–88.

Grafton, Anthony. "How Revolutionary Was the Print Revolution?" *American Historical Review* 107 (2002): 84–86.

Grafton, Anthony, April Shelford, and Nancy G. Siraisi. *New Worlds, Ancient Texts: The Power of Tradition and the Shock of Discovery.* Cambridge, MA: Harvard University Press, 1992.

Green, Daryl. "The Discovery of Another Copy of the Waldseemüller Globe (1507)." Thinking 3D, 11 December 2017. https://www.thinking3d.ac.uk /Waldseem%C3%BCller1507/.

Greenblatt, Stephen. *Marvelous Possessions: The Wonder of the New World.* Chicago: University of Chicago Press, 1991.

Grenacher, F. "The Woodcut Map. A Form-Cutter of Maps Wanders through Europe in the First Quarter of the Sixteenth Century." *Imago Mundi* 24 (1970): 31–41.

Guerreiro, Inácio. *A carta náutica de Jorge de Aguiar de 1492.* Lisbon: Academia de Marinha, 1992.

Guidi-Bruscoli, Francesco. "Bartolomeo Marchionni: Un mercador-banqueiro Florentino em Lisboa (Séculos XV–XVI)." In *Le nove son tanto e tante buone, che dir non se pò: Lisboa dos Italianos: História e Arte (Sécs. XIV–XVIII),* edited by Nunziatella Alessandre, Pedro Flor, Mariagrazia Russo, and Gaetano Sabatini, 39–60. Lisbon: Cátedra de Estudos Sefarditas "Alberto Benveniste" da Universidade de Lisboa, 2013.

Hagerty, James R. "Real-Estate Mogul Made the Deal of a Lifetime for 16th Century Map of the World." *Wall Street Journal,* 12 October 2018.

Hall, Cynthia A. "Before the Apocalypse: German Prints and Illustrated Books, 1450–1500." *Harvard University Art Museums Bulletin* 4, no. 2 (1996): 8–29.

Hamlin, Irene, and Marguerite Ragnow. "Martin Waldseemüller and the Map that Named America." Online Exhibit. James Ford Bell Library, the University of Minnesota, Minneaopolis. http://gallery.lib.umn.edu/exhibits/show/maps-and-mapmakers--martin -wal.

Hamy, E. T. *Etudes historiques et geographiques.* Paris: E. Leroux, 1896.

———. "L'oeuvre geographique des Reinel et la découverte des Moluques." In *Études historiques et héographiques,* 145–77. Paris: E. Leroux, 1896.

———. "Mecia de Viladestes: Cartographe Juif majorcain du commencement du XVe Siècle." In *Comptes Rendus des Séances de l'Academie des Inscriptions et Belles-Lettres,* 1–5. Paris: Alphose Picard et Fils, 1902.

Hardwick, Julie, Sarah M. S. Pearsall, and Karin Wulf, eds. "Centering Families in Atlantic Histories." *William and Mary Quarterly,* special issue, 70, no. 2 (2013).

Harley, J. B. "Deconstructing the Map." *Cartographia* 26 (1989): 1–20.

———. "Introduction: Texts and Contexts in the Interpretation of Early Maps," In *From Sea Charts to Satellite Images: Interpreting North American History through Maps,* edited by David Buisseret, 3–15. Chicago: University of Chicago Press, 1990.

———. "Silences and Secrecy: The Hidden Agenda of Cartography in Early Modern Europe." *Imago Mundi* 40 (1988): 57–76.

Harley, J. B., Ellen Hanlon, and Mark Warhus. *Maps and the Columbian Encounter: An Interpretive Guide to the Travelling Exhibition.* Milwaukee: Golda Meir Library, University of Wisconsin, 1990.

Harley, J. B., and Paul Laxton. *The New Nature of Maps: Essays in the History of Cartography.* Baltimore: Johns Hopkins University Press, 2001.

Harley, J. B., and David Woodward. "Concluding Remarks." In *The History of Cartography*, vol. 1: *Cartography in Prehistoric, Ancient, and Medieval Europe and the Mediterranean*, edited by J. B. Harley and David Woodward, 502–9. Chicago: University of Chicago Press, 1987.

———. *The History of Cartography*. Chicago: University of Chicago Press, 1987.

Harris, Elizabeth. "The Waldseemüller World Map: A Typographic Appraisal." *Imago Mundi* 37 (1985): 30–53.

Harrisse, Henry. *Christophe Colomb devant l'histoire*. Paris: H. Welter, 1892.

———. *Description of Works Relating to America Published between the Years 1492 and 1551*. Bibliotheca Americana Vetustissima. New York: Geo. P. Philes, 1866.

———. *The Discovery of North America: A Critical, Documentary, and Historic Investigation, with an Essay on the Early Cartography of the New World*. London: 1892; rpt., Amsterdam: N. Israel, 1969.

———. *Jean et Sébastian Cabot leur origine et leurs voyages*. Paris: E. Leroux, 1882.

———. *Les Corte-Real et leurs voyages au nouveau-monde d'après des documents nouveaux . . .* Paris: E. Leroux, 1883.

Hartigan-O'Connor, Ellen. "Marriage and Family in the Atlantic World." In *obo* in Atlantic History. https://www.oxfordbibliographies.com/view/document/obo-9780199730414/obo-9780199730414-0038.xml.

Harvey, P. D. A. *Maps in Tudor England*. Chicago: University of Chicago Press, 1993.

Harvey, P. D. A., and Scott D. Westrem. *The Hereford World Map: Mappa Mundi*. 2 vols. London: Folio Society, 2010.

Hawthorne, Walter. *From Africa to Brazil: Culture, Identity, and an Atlantic Slave Trade, 1600–1830*. New York: Cambridge University Press, 2010.

Hayes, Derek. "A Million-Dollar Map." *Fine Books and Collections*, September–October 2005. https://www2.finebooksmagazine.com/issue/0305/fine_maps.phtml.

Heawood, Edward. "A Hitherto Unknown World Map of A.D. 1506." *Geographical Journal* 62 (1923): 279–93.

Hebert, John. "The Map That Named America." *Library of Congress Information Bulletin*. 62, no. 9 (September 2003). https://www.loc.gov/loc/lcib/0309/maps.html.

Hedges, S. Blair, and Jessica K. Templeton. "Refining the Historical Timeline of the Name of America within the Print Clock." Paper presented at the Exploring Waldseemüller's World: An International Symposium, Washington, DC, 14–15 May 2009.

Herbermann, Charles George. "The Waldseemüller Map of 1507." *Historical Records and Studies* 3 (1904): 320–42.

Hess, Daniel, and Thomas Eser, eds. *The Early Dürer*. Nuremberg: Verlag des Germanischen Nationalmuseums, 2012.

Hessler, John W. "Infinite Geometries: Mathematical Notes on Werner's Commentary on Ptolemy's First Book and the Projection of the 1507 World Map by Martin Waldseemüller." Posted 15 July 2010. http://archaeolab.blogspot.com/2010/.

———. "Redrawing Waldseemüller, Writing Ptolemy: Unsolved Problems in Waldseemüller Scholarship." Paper presented at the Phillips Society Spring Meeting, Library of Congress, 16 May 2013.

———. *A Renaissance Globemaker's Toolbox: Johannes Schöner and the Revolution of Modern Science, 1475–1550*. Washington, DC: Library of Congress, 2013.

———. "Translation of Large Text Block on Sheet 9 of 1516 Carta Marina by Martin Waldseemüller." https://www.kislakfoundation.org/collections_maps.html.

———. "Warping Waldseemüller: A Phenomenological and Computational Study of the 1507 World Map." *Cartographica* 41 (2006): 101–13.

———. "What Can Waldseemüller's Projection Tell Us?" http://archaeolab.blogspot .com/2006/. Posted 15 October 2006.

Hessler, John W., Cyntia Karnes, and Heather Wanser. "Who Printed Waldseemüller: Watermark Evidence from the 1507, 1513 and 1516 Maps." *Jay I. Kislak Foundation Collections Geography and Maps* (blog). http://www.kislakfoundation.org/collections _maps.html.

Hessler, John W., and Chet Van Duzer. *Seeing the World Anew: The Radical Vision of Martin Waldseemüller's 1507 and 1516 World Maps.* Washington, DC: Library of Congress, 2012.

Hiatt, Alfred. "Mutation and Nation: The 1513 Strasbourg Ptolemy." In *Ptolemy's Geography in the Renaissance*, edited by Zur Shalev and Charles Burnett, 143–61. London: Warburg Institute, 2011.

———. *Terra Incognita: Mapping the Antipodes before 1600.* Chicago: University of Chicago Press, 2008.

Hind, Arthur Mayger. *An Introduction to a History of Woodcut: With a Detailed Survey of Work Done in the Fifteenth Century.* New York: Dover, 1963.

———. *Some Early Italian Engravings before the Time of Marcantonio.* Boston, Museum of Fine Arts, 1913.

Homer. *Odyssey.* Translated by A. T. Murray; rev. ed. by George E. Dimock. Cambridge, MA: Harvard University Press, 2014.

Horden, Peregrine, and Nicholas Purcell. "The Mediterranean and 'the New Thalassology.'" *American Historical Review* 111(2006): 722–40.

Howard, Deborah. "Venice as a Dolphin: Further Investigations into Jacopo de' Barbari's View." *Artibus et Historiae* 18, no. 35 (1997): 101–11.

Hulme, Peter. *Colonial Encounters: Europe and the Native Caribbean, 1492–1787.* London: Methuen, 1986.

Hutchison, Jane Campbell. *Early German Artists: Martin Schongauer, Ludwig Schongauer, and Copyists. The Illustrated Bartsch*, vol. 8, Commentary, Part 1. New York; Abaris Books, 1978.

Irby, Georgia L. "Mapping the World: Greek Initiatives from Homer to Eratosthenes." In *Ancient Perspectives: Maps and Their Place in Mesopotamia, Egypt, Greece, and Rome*, edited by Richard J. A. Talbert, 81–108. Chicago: The University of Chicago Press, 2012.

Jackson, Sharyn. "How a Rare Map at the U of M Brought down a $1M Christie's Auction --StarTribune.Com." *Star Tribune*, 10 January 2018.

Jacob, Christian. *The Sovereign Map: Theoretical Approaches in Cartography throughout History.* Edited by Edward H. Dahl and translated by Tom Conley. Chicago: University of Chicago Press, 2006.

Johnson, Christine R. *The German Discovery of the World: Renaissance Encounters with the Strange and Marvelous.* Charlottesville: University of Virginia Press, 2008.

Johnson, Christine R. "Renaissance German Cosmographers and the Naming of America." *Past and Present* 191 (2006): 3–43.

Johnson, Hildegard Binder. *Carta Marina: World Geography in Strassburg, 1525.* Minneapolis: University of Minnesota Press, 1963.

Kagan, Richard L., and Philip D. Morgan, eds. *Atlantic Diasporas: Jews, Conversos, and Crypto-Jews in the Age of Mercantilism, 1500–1800*. Baltimore: Johns Hopkins University Press, 2009.

Karrow, Robert W. "Centers of Map Publishing in Europe, 1472–1600." In *The History of Cartography*, vol. 3: *Cartography in the European Renaissance*, edited by David Woodward, 611–21. Chicago: University of Chicago Press, 2007.

———. *Mapmakers of the Sixteenth Century and Their Maps: Bio-bibliographies of the Cartographers of Abraham Ortelius, 1570; Based on Leo Bagrow's A. Ortelii Catalogus Cartographorum*. Chicago: Speculum Orbis Press, 1993.

Kazhdan, A. P., and Aleksandr Petrovich. *The Oxford Dictionary of Byzantium*. New York: Oxford University Press, 1991.

Kelley, J. E., Jr. "Non-Mediterranean Influences That Shaped the Atlantic in the Early Portolan Charts." *Imago Mundi* 31 (1979): 18–35.

Kimble, George H. "The 'Esmeraldo de Situ Orbis': An Early Portuguese Textbook on Cosmography and Navigation by Duarte Pacheco." *Osiris* 3 (1937): 88–102.

———. "Portuguese Policy and Its Influence on Fifteenth Century Cartography." *Geographical Review* 23 (1933): 653–59.

Klarer, Mario. "Cannibalism and Carnivalesque: Incorporation as Utopia in the Early Image of America." *New Literary History* 30 (1999): 389–410.

Klooster, Wim. *Revolutions in the Atlantic World: A Comparative History*. New York: New York University Press, 2009.

———. *Revolutions in the Atlantic World, New Edition: A Comparative History*. 2nd ed. New York: New York University Press, 2018.

Knight, Franklin W., and Peggy K. Liss. *Atlantic Port Cities: Economy, Culture, and Society in the Atlantic World, 1650–1850*. Knoxville: University of Tennessee Press, 1991.

Koeman, Cornelis. *The History of Abraham Ortelius and His Theatrum Orbis Terrarum*. New York: Elsevier, 1964.

Kulikoff, Allan. *Tobacco and Slaves: The Development of Southern Cultures in the Chesapeake, 1680–1800*. Chapel Hill: University of North Carolina Press, 1986.

Kupčík, Ivan. *Münchner Portolankarten: Kunstmann I–XIII, und zehn weitere Portolankarten / Munich Portolan Charts: Kunstmann I–XIII, and Ten Further Portolan Charts*. Munich: Deutscher Kunstverlag, 2000.

Labi, Aisha. "An American Treasure Turns Up." *Chronicle of Higher Education*, 4 July 2012. https://chronicle.com/blogs/tweed/unknown-copy-of-map-that-gave-america-its-name -turns-up-in-munich-library/29773.

Landau, David, and Peter W. Parshall. *The Renaissance Print, 1470–1550*. New Haven: Yale University Press, 1994.

Lehmann, Martin. "The Depiction of America on Martin Waldseemüller's World Map from 1507—Humanistic Geography in the Service of Political Propaganda." *Cogent Arts and Humanities* 3, no. 1 (2016). https://www.cogentoa.com/article/10.1080/23311983.2016 .1152785.

Lehmann-Haupt, Hellmut. "Book Illustration in Augsburg in the Fifteenth Century." *Metropolitan Museum Studies* 4 (1932): 3–17.

Leitão, Henrique, and Joaquim Alves Gaspar. "Globes, Rhumb Tables, and the Prehistory of the Mercator Projection." *Imago Mundi* 66 (2014): 180–95.

Leitch, Stephanie. "Burgkmair's Peoples of Africa and India (1508) and the Origins of Ethnography in Print." *Art Bulletin* 91: 2 (2009): 134–59.

———. *Mapping Ethnography in Early Modern Germany: New Worlds in Print Culture.* New York: Palgrave Macmillan, 2010.

Leite, Duarte. "O mais antigo mapa do Brasil." In *História da colonização portuguesa do Brasil*, edited by Carlos Malheiro Dias, Roque Gameiro, and Ernesto de Vasconcellos, 2:223–81. Porto: Lithografia Nacional, 1923.

Lestringant, Frank. *Cannibals: The Discovery and Representation of the Cannibal from Columbus to Jules Verne.* Translated by Rosemary Morris. Berkeley: University of California Press, 1997.

Lewis, John. *Printed Ephemera: The Changing Uses of Type and Letterforms in English and American Printing.* Ipswich: W. S. Cowell, 1962.

Lewis, Martin W. "Dividing the Ocean Sea." *Geographical Review* 89 (1999): 188–214.

"Library of Congress Acquires Only Known Copy of 1507 World Map Compiled by Martin Waldseemüller." News from the Library of Congress. Press Release 01-093, 2001-07-23. https://www.loc.gov/item/prn-01-093/.

Linte, Guillaume. "As 'novas humanidades' na *Crónica de Guiné* de Gomes Eanes de Zurara." *Revista Portuguesa de História* 46 (2015): 15–34.

Lišěák, Vladimír. "Mapa Mondi (Catalan Atlas of 1375), Majorcan Cartographic School, and 14th Century Asia." *Proceedings of the ICA* 1 (16 May 2018): 1–8.

Long, Pamela O., David McGee, and Alan M. Stahl, eds. *The Book of Michael of Rhodes: A Fifteenth-Century Maritime Manuscript.* Vol. 3: *Studies.* Cambridge, MA: MIT Press, 2009.

Lovejoy, Paul. "The Children of Slavery—the Transatlantic Phase." *Slavery and Abolition* 27 (2006): 197–217.

Macías, Luis A. Robles. "Juan de La Cosa's Projection: A Fresh Analysis of the Earliest Preserved Map of the Americas." *Coordinates: Online Journal of the Map and Geography Round Table, American Library Association,* ser. A, no. 9 (2010): 1–42.

Mahony, Mary Ann. "The Local and the Global: Internal and External Factors in the Development of Bahia's Cacao Sector." In *From Silver to Cocaine: Latin American Commodity Chains and the Building of the World Economy, 1500–2000,* edited by Steven Topik, Carlos Marichal, and Zephyr Frank, 174–203. Durham: Duke University Press, 2006.

Maia, Carlos Roma Machado de Faria e. *Apontamentos para um novo índice cronológico das primeiras viagens, descobrimentos e conquistas dos Portugueses.* Lisbon: Oficinas Fernandes, 1937.

Manning, Patrick. "Frontiers of Family Life: Early Modern Atlantic and Indian Ocean Worlds." *Modern Asian Studies* 43 (2009): 315–33.

Marcocci, Giuseppe. "Toward a History of the Portuguese Inquisition Trends in Modern Historiography (1974–2009)." *Revue de l'histoire des religions,* no. 227 (2010): 355–93.

Markley, Lia. "The New World in Renaissance Italy: A Vicarious Conquest of Art and Nature at the Medici Court." PhD diss., University of Chicago, 2008.

Marques, Alfredo Pinheiro. *A Cartografia dos descobrimentos portugueses.* Lisbon: Ediço ELO, 1994.

———. *Origem e desenvolvimento da cartografia portuguesa na época dos descobrimentos.* Lisbon: Impr. Nacional-Casa da Moeda, 1987.

Marques, Alfredo Pinheiro. "The Dating of the Oldest Portuguese Charts." *Imago Mundi* 41 (1989): 87–97.

———. *A historiografia dos descobrimentos e expansão portuguesa*. Coimbra: Minerva, 1991.

Marques, João Martins da Silva, and Alberto Iria, eds. *Descobrimentos portugueses*. Lisbon: Instituto Nacional de Investigação Científica, 1988.

Martel, H. E. "Hans Staden's Captive Soul: Identity, Imperialism, and Rumors of Cannibalism in Sixteenth-Century Brazil." *Journal of World History* 17 (2006): 51–69.

Martín-Merás, María Luisa. *Cartografía Marítima Hispana: la imagen de América*. Barcelona: Lunwerg, 1993.

———. "La Carta de Juan de La Cosa: interpretación e historia." *Monte Buciero* 4 (2000): 71–85.

Maxwell, Kenneth. "Portugal, Europe, and the Origins of the Atlantic Commercial System, 1415–1520." *Portuguese Studies* 8 (1992): 3–16.

McCorkle, Barbara B. *New England in Early Printed Maps, 1513 to 1800: An Illustrated Carto-Bibliography*. Providence: John Carter Brown Library, 2001.

McCreery, David. "Indigo Commodity Chains in the Spanish and British Empire, 1560–1860." In *From Silver to Cocaine: Latin American Commodity Chains and the Building of the World Economy, 1500–2000*, edited by Steven Topik, Carlos Marichal, and Zephyr Frank, 53–75. Durham: Duke University Press, 2006.

McGuirk, Donald, Jr. "Ruysch World Map: Census and Commentary." *Imago Mundi* 41 (2008): 133–41.

McKinstry, Richard. "Maps." Ephemera Society of America. http://www.ephemerasociety.org /examples/ex-maps.html.

McIntosh, Gregory C. *The Johannes Ruysch and Martin Waldseemüller World Maps: The Interplay and Merging of Early Sixteenth Century New World Cartographies*. Cerritos, CA: Plus Ultra Publishing, 2015.

———. *The Piri Reis Map of 1513*. Athens: University of Georgia Press, 2000.

———. *The Vesconte Maggiolo World Map of 1504 in Fano, Italy*. Cerritos, CA: Plus Ultra Publishing, 2013.

Meine, Karl Heinz. *Erläuterungen Zur Ersten Gedruckten (Strassen-) Wandkarte von Europa Der Carta Itineraria Evropae der Jahre 1511 Bzw. 1520 von Martin Waldseemüller (um 1470 bis etwa 1521): Kostbarkeit des Tiroler Landesmuseum Ferdinandeum, Innsbruck*. Bonn: Kirschbaum Verlag, 1971.

Meinig, Donald W. *The Shaping of America: A Geographical Perspective on 500 Years of History*. Vol. 1: *Atlantic America, 1492–1800*. New Haven: Yale University Press, 1986.

Melville, Elinor G. K. *A Plague of Sheep: Environmental Consequences of the Conquest of Mexico*. Cambridge: Cambridge University Press, 1997.

Menna, F., A. Rizzi, E. Nocerino, F. Remondino, and A. Gruen. "High Resolution 3D Modeling of the Behaim Globe." *International Archives of the Photogrammetry, Remote Sensing and Spatial Information Sciences* XXXIX-B5 (2012): 115–20.

Metcalf, Alida C. "Amerigo Vespucci and the Four Finger (Kunstmann II) World Map." *e-Perimetron* 7 (2012): 36–44.

———. *Go-betweens and the Colonization of Brazil, 1500–1600*. Austin: University of Texas Press, 2005.

———. "Mapping the Traveled Space: Hans Staden's Maps in *Warhaftige Historia*." *E-JPH* 7(2009): 1–15.

————. "Who Cares Who Made the Map? *La Carta del Cantino* and Its Anonymous Maker." *e-Perimetron* 12 (2017): 1–23.

Meurer, Peter H. "Cartography in the German Lands, 1450–1650." In *The History of Cartography*, vol. 3: *Cartography in the European Renaissance*, edited by David Woodward, 1172–1245. Chicago: University of Chicago Press, 2007.

Meyerson, Mark D. *The Muslims of Valencia: In the Age of Fernando and Isabel, between Coexistence and Crusade*. Berkeley: University of California Press, 1991.

Michienzi, Ingrid Houssaye, and Emmanuelle Vagnon. "Commissioning and Use of Charts Made in Majorca, c.1400: New Evidence from a Tuscan Merchant's Archive." *Imago Mundi* 71 (2018): 22–33.

Milano, Ernesto. *La Carta del Cantino*. Modena: Il Bulino, 1991.

Milano, Ernesto, and Annalisa Battini. *Mapamundi Catalán Estense; Escuela Cartográfica Mallorquina*. Barcelona: M. Moleiro Editor, S.A., 1966.

Ministério da Cultura. *A historia da cartografia na obra do 2° Visconde de Santarém*. Lisbon: Biblioteca Nacional, 2006.

Mintz, Sidney W. *Sweetness and Power: The Place of Sugar in Modern History*. New York: Viking, 1985.

Mintzker, Yair. "Between the Linguistic and the Spatial Turns: A Reconsideration of the Concept of Space and Its Role in the Early Modern Period." *Historical Reflections / Réflexions Historiques* 35 (2009): 37–51.

Miró, Mónica et al. *Atlas Miller*. Barcelona: M. Moleiro Editor, S.A., 2006.

Missinne, Stefaan. "America's Birth Certificate: The Oldest Globular World Map: C. 1507." *Advances in Historical Studies* 4 (2015): 239–307.

Mittman, Asa Simon. *Maps and Monsters in Medieval England*. New York: Routledge, 2006.

Mittman, Asa Simon, with Peter Dendle, eds. *The Ashgate Research Companion to Monsters and the Monstrous*. London: Routledge, 2013.

Mollat, Michel, and Monique de La Roncière. *Sea Charts of the Early Explorers: 13th to 17th Century*. New York: Thames and Hudson, 1984.

Monmonier, Mark S. *Coast Lines: How Mapmakers Frame the World and Chart Environmental Change*. Chicago: University of Chicago Press, 2008.

Moraes, Rubens Borba de. *Bibliographia Brasiliana*. 2 vols. Amsterdam: Colibris Editora, 1958.

Moreira, Rafael. "Pedro e Jorge Reinel (at. 1504–60): Dois cartógrafos negros na côrte de d. Manuel de Portugal (1495–1521)." *Terra Brasilis* 4 (2015) https://journals.openedition.org /terrabrasilis/1209.

Mota, A. Teixeira da. *A Evolução da ciência náutica durante os séculos XV–XVI na cartografia portuguesa da época*. Lisbon: Junta de Investigações do Ultramar, 1961.

Nebenzahl, Kenneth. *Atlas of Columbus and the Great Discoveries*. Chicago: Rand McNally, 1990.

————. *Mapping the Silk Road and Beyond: 2000 Years of Exploring the East*. London: Phaidon Press, 2004.

Nicolai, Roel. *The Enigma of the Origin of Portolan Charts: A Geodetic Analysis of the Hypothesis of a Medieval Origin*. Leiden: Brill, 2016.

Norton, F. J. *A Descriptive Catalogue of Printing in Spain and Portugal, 1501–1520*. Cambridge: Cambridge University Press, 1978.

Norton, Marcy. *Sacred Gifts, Profane Pleasures: A History of Tobacco and Chocolate in the Atlantic World.* Ithaca: Cornell University Press, 2010.

Nunn, George E. *The Mappemonde of Juan de La Cosa: A Critical Investigation of Its Date.* Jenkintown, PA: George H. Beans Library, 1934.

Nuti, Lucia. "The Perspective Plan in the Sixteenth Century: The Invention of a Representational Language." *Art Bulletin* 76, no. 1 (March 1994): 105–28.

Obhof, Ute. "The Terrestrial Globe That Named America: On Historical Preservation of Martin Waldseemüller's Gores." *Globe Studies*, nos. 55–56 (2009): 13–21.

Ohl des Marais, Albert. *Mathias Ringmann dit "Philesius," graveur en bois, par . . .* Saint-Dié: impr. Cuny, 1933.

Olival, Fernanda. "The Military Orders and the Nobility in Portugal, 1500–1800." *Mediterranean Studies* 11 (2002): 71–88.

Oliveira, Guarino Alves D'. *A costa setentrional do Brasil na Carta de Navegar de Alberto Cantino.* [Ceará]: [A Fortaleza], 1968.

Padrón, Ricardo. "Mapping Plus Ultra: Cartography, Space, and Hispanic Modernity." *Representations* 79 (2002): 28–60.

Palencia-Roth, Michael. "Cannibalism and the New Man of Latin America in the 15th- and 16th-Century European Imagination." *Comparative Civilizations Review* 12, no. 12 (1985): 1–27.

———. "Mapping the Caribbean: Cartography and the Cannibalization of Culture." In *A History of Literature in the Caribbean*, edited by A. James Arnold, 3:3–28. Amsterdam: John Benjamins, 1997.

Panofsky, Erwin. *Albrecht Dürer.* 2 vols. 3rd ed. Princeton: Princeton University Press, 1948.

Paquette, Gabriel. *Imperial Portugal in the Age of Atlantic Revolutions.* Cambridge: Cambridge University Press, 2014.

Paullin, Charles Oscar, and John Kirtland Wright. *Atlas of the Historical Geography of the United States.* Washington, DC, and New York: Carnegie Institution of Washington and the American Geographical Society, 1932.

Penfold, Peter A. *Maps and Plans in the Public Record Office.* London: Her Majesty's Stationery Office, 1967.

Penney, Clara Louisa. *Printed Books, 1468–1700, in the Hispanic Society of America.* New York: Hispanic Society of America, 1965.

Pereira, José Manuel Malhão. "A ciência náutica e o contacto entre os povos." Unpublished paper.

Pereira, Moacyr Soares. *A navegação de 1501 ao Brasil e Américo Vespúcio.* Rio de Janeiro: Artes Gráficas Ltda, 1984.

———. "O novo mundo no planisfério da Casa de Este, o 'Cantino.'" *Revista da Universidade de Coimbra* 35 (1989): 271–308.

Pérez, Sandra Sáenz-López. *The Beatus Maps: The Revelation of the World in the Middle Ages.* Translated by Peter Krakenberger and Gerry Coldham. Burgos: Siloé, 2014.

———. "La Carta de Juan de la Cosa (1500), colofón de la cartografía medieval." *Piezas del Mes,* Museo Naval de Madrid 2003/2005, 10–31.

———. "La pluralidad religiosa del mundo en el siglo XV a través de la carta náutica de Mecia de Viladestes (1413)." *Anales de Historia del Arte* 22 (2012): 389–404.

———. "Medieval Imagery and Knowledge of the World in Spanish Cartography." *Imago Mundi* 58 (2006): 238–39.

Pettegree, Andrew. *The Book in the Renaissance*. New Haven: Yale University Press, 2010.

Plummer, Marjorie Elizabeth, and Robin B. Charles, eds. *Ideas and Cultural Margins in Early Modern Germany: Essays in Honor of H. C. Erik Midelfort*. Aldershot, UK: Ashgate, 2009.

Pohl, Frederick Julius. *Amerigo Vespucci, Pilot Major*. New York: Columbia University Press, 1944.

Prestage, Edgar. *The Portuguese Pioneers*. New York: Barnes and Noble 1967.

Proverbio, Edoardo. "Astronomical and Sailing Tables from the Second Half of the 15th Century to the Middle of the 16th Century." *Società Astronomica Italiana* 64 (1991): 469–96.

Puchades i Bataller, Ramon J. *Les cartes portolanes: La representació medieval d'una mar solcada*. Translated by Richard Rees. Barcelona: Institut Cartogràfic de Catalunya, 2007.

Quinn, David. "Artists and Illustrators in the Early Mapping of North America." *Mariner's Mirror* 72, no. 3 (1986): 244–73.

Ragnow, Marguerite. "Columbus' Maps: European and Arab Maps in Circulation." Paper presented at the American Historical Association, San Diego, January 2010.

Randles, W. G. L. "From the Mediterranean Portulan Chart to the Marine World Chart of the Great Discoveries: The Crisis in Cartography in the Sixteenth Century." *Imago Mundi* 40 (1988): 115–18.

Ratti, Antonio. "A Lost Map of Fra Mauro Found in a Sixteenth Century Copy." *Imago Mundi* 40 (1988): 77–85.

Rau, Virgínia. *O açúcar da Madeira nos fins do século XV: Problemas de produção e comércio*. Lisbon: Junta-Geral do Distrito Autónomo de Funchal, 1962.

Ravenstein, Ernest George. *Martin Behaim, His Life and His Globe*. London: G. Philip & Son, 1908.

———. "The Voyages of Diogo Cão and Bartholomeu Dias, 1482–88." *Geographical Journal* 16 (1900): 625–55.

Reed, Ronald. *The Nature and Making of Parchment*. Leeds: Elmete, 1975.

Relaño, Francesc. *The Shaping of Africa: Cosmographic Discourse and Cartographic Science in Late Medieval and Early Modern Europe*. Aldershot, UK: Ashgate, 2002.

Restall, Matthew. *Seven Myths of the Spanish Conquest*. Oxford: Oxford University Press, 2004.

Ricci, Milena. *Carte geografiche miniate delle collezioni estense*. Translated by Paulo Zanazzo. Modena: Biblioteca estense universitaria, 2011.

Richardson, William A. R. "South America on Maps before Columbus? Martellus's 'Dragon's Tail' Peninsula." *Imago Mundi* 55 (2003): 25–37.

Rickards, Maurice, and Michael Twyman, eds. *The Encyclopedia of Ephemera: A Guide to the Fragmentary Documents of Everyday Life for the Collector, Curator, and Historian*. New York: Routledge, 2000.

Ristow, Walter W., and R. A. Skelton. *Nautical Charts on Vellum in the Library of Congress*. Washington, DC: Library of Congress, 1977.

Robinson, Arthur H. "Mapmaking and Map Printing: The Evolution of a Working Relationship." In *Five Centuries of Map Printing*, edited by David Woodward, 1–23. Chicago: University of Chicago Press, 1975.

Rodger, N. A. M. "Atlantic Seafaring." In *The Oxford Handbook of the Atlantic World, 1450–1850*, edited by Nicolas Canny and Philip Morgan, 71–86. Oxford: Oxford University Press, 2011.

Rodney, Walter. *A History of the Upper Guinea Coast, 1545–1800*. Oxford Studies in African Affairs. Oxford: Clarendon Press, 1970.

Romm, James S. *The Edges of the Earth in Ancient Thought: Geography, Exploration, and Fiction*. Princeton: Princeton University Press, 1992.

Ronsin, Albert. "Hugues des hazards et le Gymnase Vosgien de Saint-Die." *Annales de l'Est* 2 (2005): 151–65.

Rosen, Mark. *The Mapping of Power in Renaissance Italy: Painted Cartographic Cycles in Social and Intellectual Context*. New York: Cambridge University Press, 2015.

Roukema, E. "Brazil in the Cantino Map." *Imago Mundi* 17 (1963): 7–26.

———. "Some Remarks on the La Cosa Map." *Imago Mundi* 14 (1959): 38–54.

Rubiés, Joan Pau. "Texts, Images and the Perception of 'Savages' in Early Modern Europe: What We Can Learn from White and Harriot." In *European Visions: American Voices*, edited by Kim Sloan, 120–30. London: British Museum Research Publications, 2009.

———. "The Worlds of Europeans, Africans, and Americans, c. 1490." In *The Oxford Handbook of the Atlantic World, 1450–1850*, edited by Nicolas Canny and Philip Morgan, 21–37. Oxford: Oxford University Press, 2011.

Ruhstaller, Stefan. "Bartolomé de las Casas y su copia del 'Diario de a Bordo' de Colón. Tipología de las apostillas." *Cauce Revista Internacional de Filología, Comunicación y sus Didácticas* 14–15 (1992): 615–37.

Russell, P. E. *Prince Henry "the Navigator": A Life*. New Haven: Yale University Press, 2000.

Sánchez, Antonio. *La espada, la cruz y el Padrón: soberanía, fe y representación cartográfica en el mundo ibérico bajo la Monarquía Hispánica, 1503–1598*. Madrid: Consejo Superior de Investigaciones Científicas, 2013.

Sander, Jochen, ed. *Albrecht Dürer: His Art in Context*. Munich: Prestel, 2013.

Sandler, Martin W., and Dennis Reinhartz. *Atlantic Ocean: The Illustrated History of the Ocean That Changed the World*. New York: Sterling, 2008.

Sandman, Alison. "An Apologia for the Pilots' Charts: Politics, Projections and Pilots' Reports in Early Modern Spain." *Imago Mundi* 56 (2004): 7–22.

———. "Cosmographers vs. Pilots: Navigation, Cosmography, and the State in Early Modern Spain." PhD diss., University of Wisconsin at Madison, 2001.

———. "Spanish Nautical Cartography in the Renaissance." In *The History of Cartography*, vol. 3: *Cartography in the European Renaissance*, edited by David Woodward, 1095–1142. Chicago: University of Chicago Press, 2007.

Sandys, Sir John, trans. *The Odes of Pindar*. Cambridge, MA and London: Cambridge University Press and William Heinemann, 1968.

Santarém, Manuel Francisco de Barros e Sousa. *Essai sur l'histoire de la cosmographie et de la cartographie*. Paris: Imprimerie Maulde et Renou, 1849.

———. *Inéditos (Miscellanea)*. Lisbon: L. da Silva, 1914.

———. *Opusculos e esparsos*. 2 vols. Lisbon: L. da Silva, 1910.

Sarson, Steven. *The Tobacco-Plantation South in the Early American Atlantic World*. London: Palgrave Macmillan, 2013.

Sauer, Carl Ortwin. *The Early Spanish Main*. Berkeley: University of California Press, 1966.

Scafi, Alessandro. *Mapping Paradise: A History of Heaven on Earth*. London: British Library, 2006.

———. "Mapping the End: The Apocalypse in Medieval Cartography." *Literature and Theology* 26 (2012): 400–416.

Schilder, Günter. *Early Dutch Maritime Cartography*. Leiden: Brill; Hes and De Graaf, 2017.

———. *Monumenta cartographica Neerlandica*. Alphen aan den Rijn: Uitg. Canaletto, 2000.

———. *Sailing for the East: History and Catalogue of Manuscript Charts on Vellum of the Dutch East India Company (VOC), 1602–1799*. Houten, Netherlands: Hes and De Graaf, 2010.

Schilder, Günter, and Hans Kok. *Sailing for the East: History and Catalogue of Manuscript Charts on Vellum of the Dutch East India Company (VOC), 1602–1799*. Utrecht Studies of the History of Cartography, 10; Houten, Netherlands: Hes and De Graaf, 2010.

Schorsch, Jonathan. *Swimming the Christian Atlantic: Judeoconversos, Afroiberians and Amerindians in the Seventeenth Century*. Leiden: Brill, 2009.

Scheffler, Michael J. "Vespucci Rediscovers America: The Pictorial Rhetoric of Cannibalism in Early Modern Culture." *Art History* 28 (2005): 295–310.

Schuller, Rudolph. "The Oldest Known Illustration of South American Indians." *Journal de La Société des Américanistes* 16 (1924): 111–18.

Schulz, Juergen. "Jacopo de' Barbari's View of Venice: Map Making, City Views, and Moralized Geography before the Year 1500." *Art Bulletin* 60: 3 (1978): 425–74.

———. "Maps as Metaphors: Mural Map Cycles of the Italian Renaissance." In *Art and Cartography: Six Historical Essays*, edited by David Woodward, 97–122. Chicago: University of Chicago Press, 1987.

Schwartz, Seymour I., and Ralph E. Ehrenberg. *The Mapping of America*. New York: H. N. Abrams, 1980.

Schwartz, Stuart B. *All Can Be Saved: Religious Tolerance and Salvation in the Iberian Atlantic World*. New Haven: Yale University Press, 2009.

———. *Sugar Plantations in the Formation of Brazilian Society: Bahia, 1550–1835*. Cambridge: Cambridge University Press, 1985.

———, ed. *Tropical Babylons: Sugar and the Making of the Atlantic World, 1450–1680*. Chapel Hill: University of North Carolina Press, 2004.

Schweitzer, Frederick M., and Harry E. Wedeck. *Dictionary of the Renaissance*. New York: Philosophical Library, 1967.

Scott, Rebecca J., and Jean M. Hébrard. *Freedom Papers: An Atlantic Odyssey in the Age of Emancipation*. Cambridge MA: Harvard University Press, 2012.

Seed, Patricia, and German Diaz. "Rewriting the History of Mapping." *EdUC Proceeding Abstract*, 2004.

Sensbach, Jon F. *Rebecca's Revival: Creating Black Christianity in the Atlantic World*. Cambridge, MA: Harvard University Press, 2006.

Servicio Histórico Militar. *Cartografía y Relaciones Históricas de Ultramar*. 9 vols. Madrid: Servicio Histórico Militar, 1983.

Shalev, Zur, and Charles Burnett, eds. *Ptolemy's Geography in the Renaissance*. London: Warburg Institute, 2011.

Sheehan, Kevin E. "The Functions of Portolan Maps: An Evaluation of the Utility of Manuscript Nautical Cartography from the Thirteenth through Sixteenth Centuries." PhD diss., Durham University, 2014.

Shirley, Rodney W. *The Mapping of the World: Early Printed World Maps, 1472–1700.* London: Holland Press, 1983; 3rd ed., London: New Holland, 1993.

Shone, Richard, and John-Paul Stonard. *The Books That Shaped Art History: From Gombrich and Greenberg to Alpers and Krauss.* London: Thames & Hudson, 2013.

Short, John R. *The World through Maps: A History of Cartography.* Toronto: Firefly Books, 2003.

Sicilia, Francesco, and Mauro Bini. *Alla scoperta del mondo: l'arte della cartografia da Tolomeo a Mercatore.* Modena: Il Bulino, 2002.

Sidbury, James. *Becoming African in America: Race and Nation in the Early Black Atlantic.* Oxford: Oxford University Press, 2007.

Siegert, Bernhard. "The Map Is the Territory." *Radical Philosophy* 169 (2011): 12–16.

Silió, Fernando Cervera. *La Carta de Juan de la Cosa: Analysis cartográfico.* Santander, Spain: Fundación Marcelino Botín, 1995.

Skelton, R. A. "A Contract for World Maps at Barcelona, 1399–1400." *Imago Mundi* 22 (1968): 107–13.

Slave Voyages. https://www.slavevoyages.org/.

Smallwood, Stephanie E. *Saltwater Slavery: A Middle Passage from Africa to American Diaspora.* Cambridge, MA: Harvard University Press, 2008.

Smith, Catherine Delano. "Map Ownership in Sixteenth-Century Cambridge: The Evidence of Probate Inventories." *Imago Mundi* 47 (1995): 67–93.

———. "Signs on Printed Topographical Maps, ca. 1470–ca. 1640." In *History of Cartography,* vol. 3: *Cartography in the European Renaissance,* , edited by David Woodward, 528–90. Chicago: University of Chicago Press, 2007.

Smith, Pamela H., and Paula Findlen, eds. *Merchants and Marvels: Commerce, Science and Art in Early Modern Europe.* New York: Routledge, 2002.

Smith, Paul J. "Rereading Dürer's Representations of the Fall of Man." In *Zoology in Early Modern Culture: Intersections of Science, Theology, Philology, and Political and Religious Education,* 301–28. Leiden: Brill, 2014.

Smith, S. D. "The Atlantic History Paradigm." Review of *Atlantic History,* by Bernard Bailyn. *New England Quarterly* 79, no. 1 (2006): 123–33.

Snyder, John Parr. *Flattening the Earth: Two Thousand Years of Map Projections.* Chicago: University of Chicago Press, 1993.

Sociedade Brasileira de Cartografia. *Cartografia Portuguêsa: Roteiro de Glórias.* Rio de Janeiro: Sociedade de Cartografia, 1963.

Soucek, Svatopluk. *Piri Reis and Turkish Mapmaking after Columbus: The Khalili Portolan Atlas.* London: Nour Foundation, Aximuth Editions, and Oxford University Press, 1996.

Soule, Emily Berquist. "From Africa to the Ocean Sea: Atlantic Slavery in the Origins of the Spanish Empire." *Atlantic Studies* 51 (2018): 16–39.

Stevens, Henry, Jr. *Stevens's American Bibliographer.* Chiswick, UK: C. Whittingham, 1854.

Stevens, Wesley M. "The Figure of the Earth in Isidore's 'De Natura Rerum.'" *Isis* 71, no. 2 (1980): 268–77.

Stevenson, Edward Luther. *Terrestrial and Celestial Globes: Their History and Construction, Including a Consideration of Their Value as Aids in the Study of Geography and Astronomy.* Vol. 2. New Haven: Hispanic Society of America and Yale University Press, 1921.

Stewart, Alison G. *Before Bruegel: Sebald Beham and the Origins of Peasant Festival Imagery.* Aldershot, UK : Ashgate, 2008.

Strauss, Walter L., and Adam von Bartsch. *The Illustrated Bartsch.* New York: Abaris Books, 1978.

Strickland, Debra Higgs. "The Sartorial Monsters of *Herzog Ernst.*" *Different Visions: A Journal of New Perspectives on Medieval Art* 2 (2010): 1–35.

Studnicki-Gizbert, Daviken. *A Nation upon the Ocean Sea: Portugal's Atlantic Diaspora and the Crisis of the Spanish Empire, 1492–1640.* Oxford: Oxford University Press, 2007.

Subrahmanyam, Sanjay. *The Career and Legend of Vasco da Gama.* Cambridge: Cambridge University Press, 1997.

Sullivan, Margaret A. "The Witches of Dürer and Hans Baldung Grien." *Renaissance Quarterly* 53 (2000): 333–401.

Sweet, James H. *Domingos Álvares, African Healing, and the Intellectual History of the Atlantic World.* Chapel Hill: University of North Carolina Press, 2013.

———. *Recreating Africa: Culture, Kinship, and Religion in the African-Portuguese World, 1441–1770.* Chapel Hill: University of North Carolina Press, 2003.

Symcox, Geoffrey, Luciano Formisano, Theodore J. Cachey, and John C McLucas. *Italian Reports on America, 1493–1522: Accounts by Contemporary Observers.* Repertorium Columbianum, vol. 12. Turnhout: Brepols, 2002.

Talbert, Richard J. A. *Ancient Perspectives: Maps and Their Place in Mesopotamia, Egypt, Greece, and Rome.* Chicago: University of Chicago Press, 2012.

Tavárez, David, ed. *Words and Worlds Turned Around: Indigenous Christianities in Colonial Latin America.* Boulder: University Press of Colorado, 2017.

Taylor, Andrew. *The World of Gerard Mercator: The Mapmaker Who Revolutionized Geography.* New York: Walker, 2004.

Taylor, E. G. R. "Jean Rotz: His Neglected Treatise on Nautical Science." *Geographical Journal* 73, no. 5 (1929): 455–59.

Tenreiro, Francisco José. *A ilha de São Tomé.* Lisbon: Junta de Investigações do Ultramar, 1961.

Thornton, John. *Africa and Africans in the Making of the Atlantic World.* Cambridge: Cambridge University Press, 1998.

Thomas, Hugh. *The Slave Trade: The Story of the Atlantic Slave Trade, 1440 - 1870.* New York: Simon and Schuster, 1997.

Topik, Steven, Carlos Marichal, and Zephyr L. Frank, Eds. *From Silver to Cocaine: Latin American Commodity Chains and the Building of the World Economy, 1500–2000.* Durham: Duke University Press, 2006.

Topik, Steven and Mario Samper "The Latin American Coffee Commodity Chain: Brazil and Costa Rica." In *From Silver to Cocaine: Latin American Commodity Chains and the Building of the World Economy, 1500–2000,* edited by Steven Topik, Carlos Marichal, and Zephyr Frank, 53–75. Durham: Duke University Press, 2006.

Tuohy, Thomas. *Herculean Ferrara: Ercole d'Este, 1471–1505, and the Invention of a Ducal Capital.* Cambridge: Cambridge University Press, 1996.

Turnbull, David. "Cartography and Science in Early Modern Europe: Mapping the Construction of Knowledge Spaces." *Imago Mundi* 48 (1996): 5–24.

Unger, Richard W. *Ships on Maps: Pictures of Power in Renaissance Europe.* New York: Palgrave Macmillan, 2010.

Usque, Samuel. *Consolaçam ás tribulaçoens de Israel*. Coimbra, Portugal: França Amado, 1906.

Valente, José. "The New Frontier: The Role of the Knights Templar in the Establishment of Portugal as an Independent Kingdom." *Mediterranean Studies* 7 (1998): 49–65.

Vals, Ana Matilla, and Fernández, Manuel Mortari, eds. *Juan de la Cosa y la época de los descubrimientos*. Madrid: Sociedad Estatal de Conmemoraciones Culturales, 2010.

Van Deusen, Nancy. *Global Indios: The Indigenous Struggle for Justice in Sixteenth-Century Spain*. Durham: Duke University Press, 2015.

Van Duzer, Chet. "Details, Date, and Significance of the Fifth Set of Waldseemüller's Globe Gores Recently Discovered in the Munich University Library." Available on Academia.edu. https://www.academia.edu/1829430/Details_Date_and_Significance_of _the_Fifth_Set_of_Waldseem%C3%BCller_s_Globe_Gores_Recently_Discovered_in _the_Munich_University_Library.

———. "Graphic Record of a Lost Wall Map of the World (c. 1490) by Henricus Martellus." *Peregrinations: Journal of Medieval Art and Architecture* 5, no. 2 (2015): 48–64.

———. *Henricus Martellus's World Map at Yale (c. 1491): Multispectral Imaging, Sources, and Influence*. Cham, Switzerland: Springer, 2018.

———. "Multispectral Imaging for the Study of Historic Maps: The Example of Henricus Martellus's World Map at Yale." *Imago Mundi* 68 (2016): 62–66.

———. "A Newly Discovered Fourth Exemplar of Francesco Rosselli's Oval Planisphere of c. 1508." *Imago Mundi* 60 (2008): 195–201.

———. "A Northern Refuge of the Monstrous Races: Asia on Waldseemüller's 1516 'Carta Marina.'" *Imago Mundi* 62 (2010): 221–31.

———. *Sea Monsters on Medieval and Renaissance Maps*. London: The British Library, 2013.

———. "Waldseemüller's World Maps of 1507 and 1516: Sources and Development of His Cartographical Thought,." *The Portolan*, 85 (Winter 2012): 8–20.

Varela Marcos, Jesús, ed. *Juan de la Cosa: La cartografía histórica de los descubrimientos españoles*. Seville: Universidad Internacional de Andalucía, 2011.

Vasconcelos, Ernesto Julio de Carvalho E. *Subsidios para a historia de cartografia portugueza nos séculos XVI, XVII e XVIII*. Lisbon: Tipografia universal, 1916.

Verner, Coolie. "Copperplate Printing." In *Five Centuries of Map Printing*, edited by David Woodward, 51–75. Chicago: University of Chicago Press, 1975.

Vidal, Cécile. "For a Comprehensive History of the Atlantic World or Histories Connected in and beyond the Atlantic World?" Translated by Michèle R. Greer. *Annales. Histoire, Sciences Sociales*, vol. 67th year, no. 2(2012): 279–300.

Vignolo, Paolo. "Hic Sunt Canibales: El Canibalismo del nuevo mundo en el imaginario europeo (1492–1729)." *Anuario Colombiano de Historia Social y de La Cultura* 32 (2005): 151–188.

Vilches, Elvira. "El Atlántico en la historiografía indiana del siglo XVI." *Revista Iberoamericana* 75 (2009): 639–55.

Villas Bôas, Luciana. "The Anatomy of Cannibalism: Religious Vocabulary and Ethnographic Writing in the Sixteenth Century." *Studies in Travel Writing* 12 (2008): 7–27.

Vinson, Donna A. "The Western Sea: Atlantic History before Columbus." *Northern Mariner* 10, no. 3 (2000): 1–14.

Viterbo, Sousa. "A Pesca do Coral No Seculo XV," *Arquivo Histórico Portuguez* 1 (1903): 315–320.

———. *Trabalhos náuticos dos portugueses: séculos XVI e XVII.* Lisbon: Impr. Nacional-Casa da Moeda, 1988.

———. *Trabalhos nauticos dos portuguezes nos seculos XVI e XVII.* Lisbon: Academia Real das Sciencias, 1898.

Wagner, Henry R. "The Manuscript Atlases of Battista Agnese." In *The Papers of the Bibliographical Society of America*, 25:1–110. Chicago: University of Chicago Press, 1931.

Waldseemüller, Martin, Joseph Fischer, and Franz Wieser. *Die Älteste Karte mit dem Namen Amerika aus dem Jahre 1507 und die Carta Marina aus dem Jahre 1516 des M. Waldseemüller (Ilacomilus).* Translated by George Jacob Pickel. Innsbruck and London: Wagner and H. Stevens, Son; Stiles, 1903.

Walker, James. "From Alterity to Allegory: Depictions of Cannibalism on Early European Maps of the New World." Occasional Paper Series, 9. Washington, DC: Philip Lee Phillips Society, 2015.

Wallis, Helen. "The Rotz Atlas: A Royal Presentation." *Map Collector* 20 (September 1983): 39–42.

Wanser, Heather. "Treatment and Preparation of Waldseemüller's Map." *Information Bulletin,* September 2003. Library of Congress, Washington, DC. https://www.loc.gov /loc/lcib/0309/conserve.html.

Waters, David W. *Science and the Techniques of Navigation in the Renaissance.* London: National Maritime Museum, 1976.

Watts, Pauline Moffitt. "Prophecy and Discovery: On the Spiritual Origins of Christopher Columbus's 'Enterprise of the Indies.'" *American Historical Review* 90 (1985): 73–102.

Webster, Jane. "Slave Ships and Maritime Archaeology: An Overview." *International Journal of Historical Archaeology* 12 (2008): 6–19.

Weil-Garris, Kathleen, and John F. D'Amico. "The Renaissance Cardinal's Ideal Palace: A Chapter from *Cortesi's de Cardinalatu.*" *Memoirs of the American Academy in Rome* 35, no. 1 (1980): 45–123.

Weinstein, Donald. *Ambassador from Venice, Pietro Pasqualigo in Lisbon, 1501.* Minneapolis: University of Minnesota Press, 1960.

Weiss, Benjamin. "The *Geography* in Print: 1475–1530." In *Ptolemy's Geography in the Renaissance,* edited by Zur Shalev and Charles Burnett, 91–120. London: Warburg Institute, 2011.

Wey Gómez, Nicolás. *The Tropics of Empire: Why Columbus Sailed South to the Indies.* Cambridge, MA: MIT Press, 2008.

Wheat, David. *Atlantic Africa and the Spanish Caribbean, 1570–1640.* Chapel Hill: Omohundro Institute of Early American History and Culture, 2016.

Whitehead, Neil L. "Carib Cannibalism: The Historical Debate. *Journal de la société des américanistes* 70 (1984): 69–87.

———. "Hans Staden and the Cultural Politics of Cannibalism." *Hispanic American Historical Review* 80 (2000): 721–51.

Whitfield, Peter. *New Found Lands: Maps in the History of Exploration.* New York: Routledge, 1998.

Wigen, Kären. "Introduction: AHR Forum Oceans of History." *American Historical Review* 111, no. 3 (2006): 717–21.

Williams, Wes. "'L'Humanité du tout perdue?': Early Modern Monsters, Cannibals and Human Souls." *History and Anthropology* 23 (2012): 235–56.

Willingham, Elizabeth Moore. *The Mythical Indies and Columbus's Apocalyptic Letter.* Eastbourne: Sussex Academic Press, 2015.

Winchester, Simon. *Atlantic: Great Sea Battles, Heroic Discoveries, Titanic Storms, and a Vast Ocean of a Million Stories.* New York: Harper, 2010.

Winsor, Justin, ed. *Spanish Explorations and Settlements in North America from the Fifteenth to the Seventeenth Century: Narrative and Critical History of America.* Vol. 2. New York: Houghton, Mifflin, 1886.

Winter, Heinrich. "Catalan Portolan Maps and Their Place in the Total View of Cartographic Development." *Imago Mundi* 11 (1954): 1–12.

———. "A Late Portolan Chart at Madrid and Late Portolan Charts in General." *Imago Mundi* 7 (1950): 37–46.

———. "On the Real and the Pseudo-Pilestrina Maps and Other Early Portuguese Maps in Munich." *Imago Mundi* 4 (1947): 25–27.

———. "The Origin of the Sea Chart." *Imago Mundi* 13 (1956): 39–44.

Wolff, Hans. *America: Early Maps of the World.* Munich: Prestel, 1992.

Wood, Denis, with John Fels. *The Power of Maps.* New York: Guilford Press, 1992.

Wood, Denis, with John Fels and John Krygier,. *Rethinking the Power of Maps.* New York: Guilford Press, 2010.

Woodward, David. *Art and Cartography: Six Historical Essays.* Chicago: University of Chicago Press, 1987.

———. *Five Centuries of Map Printing.* Chicago: University of Chicago Press, 1975.

———. "The Italian Map Trade, 1480–1650." In *The History of Cartography*, vol. 3: *Cartography in the European Renaissance*, edited by David Woodward, 773–803. Chicago: University of Chicago Press, 2007.

———. "Medieval *Mappaemundi*." In *The History of Cartography*, I: *Cartography in Prehistoric, Ancient, Medieval Europe and the Mediterranean*, edited by J. B. Harley and David Woodward, 286–370. Chicago: University of Chicago Press, 1987.

———. "Reality, Symbolism, Time, and Space in Medieval World Maps." *Annals of the Association of American Geographers* 75 (1985): 510–21.

———. "Techniques of Map Engraving, Printing, and Coloring in the European Renaissance." In *The History of Cartography*, vol. 3: *Cartography in the European Renaissance*, edited by David Woodward, 591–610. Chicago: University of Chicago Press, 2007.

———. "The Woodcut Technique." In *Five Centuries of Map Printing*, edited by David Woodward, 25–50. Chicago: University of Chicago Press, 1975.

World Heritage List, Ferrara, No. 733. 20 October 1994.

"World Heritage List, Ferrara, Works of Art in Italy: Losses and Survival in War." London: His Majesty's Stationery Office, 1945.

Zamora, Margarita. *Reading Columbus.* Berkeley: University of California Press, 1993.

Zeyl, Donald, and Barbara Sattler. "Plato's Timaeus." In *The Stanford Encyclopedia of Philosophy*, edited by Edward N. Zalta. Stanford: Metaphysics Research Lab, Stanford University, 2019.

Zika, C. "Cannibalism and Witchcraft in Early Modern Europe: Reading the Visual Images." *History Workshop Journal* 44 (1997): 77–105.

Page numbers in *italics* refer to figures.